T. EASTON

DOW C[OLLEGE]
4 MAY 1989
BARRY LIBRARY

Gasoline and Diesel Fuel Additives

Critical Reports on Applied Chemistry: Editorial Board

C. A. Finch	H. J. Cottrell	J. M. Sidwick
Chairman	C. R. Ganellin, FRS	J. C. Weeks
A. R. Burkin	E. G. Hancock	
N. Blakebrough	K. R. Payne	

Critical Reports on Applied Chemistry Volume 25

Gasoline and Diesel Fuel Additives

Edited by K. Owen

Published on behalf of the Society of Chemical Industry by
John Wiley & Sons
Chichester · New York · Brisbane · Toronto · Singapore

Copyright © 1989 by The Society of Chemical Industry.

All rights reserved.

No part of this book may be reproduced by any means, or transmitted, or translated into a machine language without the written permission of the publisher

British Library Cataloguing in Publication Data information available

ISBN 0 471 92216 1

Phototypesetting by Thomson Press (India) Ltd., New Delhi and
Printed in Great Britain by Anchor Press, Tiptree, Essex.

Contents

Editor's introduction ix

1 Precombustion gasoline additives 1
K. Owen and R. G. M. Landells

2 Gasoline additives influencing combustion processes 39
D. R. Blackmore

3 Additives influencing diesel fuel combustion 65
T. J. Russell

4 Diesel fuel additives influencing flow and storage properties 105
T. R. Coley

5 The use of oxygenates in motor gasolines 133
G. J. Lang and F. H. Palmer

Index 169

List of contributors

David R. Blackmore Shell Research Ltd, Thornton Research Centre, PO Box 1, Chester CH1 3SH, England

Trevor R. Coley 5 Wheatcroft Close, Abingdon, Oxon. OX14 2BE, England

Robin G. M. Landells Lubrizol Limited, Waldron House, 57–63 Old Church Street, London SW3 5BS, England

Gregory J. Lang BP Research Centre, Chertsey Road, Sunbury-on-Thames, Middlesex TW16 7LN, England

Keith Owen Middle Farm, Church Lane, Brightwell-cum-Sotwell, Wallingford OX10 0SD, England

Francis H. Palmer BP Research Centre, Chertsey Road, Sunbury-on-Thames, Middlesex TW16 7LN, England

Trevor J. Russell The Associated Octel Company Ltd, Engine Laboratory, Watling Street, Milton Keynes MK1 1EZ, England

Editor's introduction

Gasoline and diesel fuel additives are more important today than they have ever been. This is because there has been a very significant change in the composition of both gasoline and diesel fuel over the past ten years resulting from the phasing out of lead alkyls from gasoline and the depression of the fuel oil market. Both gasoline and diesel fuel now contain more cracked components than in the past, which means that the olefin content of these fuels has increased. Olefins are less stable to oxidation and hence more likely to form deposits in a vehicle fuel system than are the paraffinic and aromatic components. They are also poorer in cetane quality. The fuels therefore need additives in order to overcome these deficiencies. Oxygenated components are also now widely used in gasoline as an extender and to compensate for the loss of lead alkyls. This can cause a number of difficulties which can be overcome by use of appropriate additives.

Another factor which has increased the importance of additives is the pressure on vehicle manufacturers to improve exhaust gas quality and fuel economy. Deposit formation can steadily worsen an engine's performance in these respects, and many new models are particularly sensitive to deposits in the inlet system. Additives can help to maintain engine performance by preventing deposit build-up.

A third factor is the wish for product differentiation on the part of oil companies, who need to demonstrate that their gasoline or diesel fuel has special features that makes it more desirable than competitive materials. The widespread practice of exchanging fuels between companies to minimize distribution costs means that the use of additives is often the only way that a company can achieve this product differentiation, since such additives can be injected into the road tanker during loading prior to delivery to the service station. There are a large number of examples of the use of additives to provide a strong basis for advertising 'new' gasolines, and the potential for market share improvement by this means has been demonstrated many times.

The aim of the book is to provide a reference work useful to three types of user:

1. Oil company employees involved in gasoline and diesel fuel marketing, refining, product quality and distribution;
2. Fuel additive manufacturing companies, to provide them with a comprehensive review of user requirements and concerns as well as a general reference manual;
3. Automotive engineers who are concerned with fuel systems, to help in their understanding of the benefits that additives can bestow and the problems that improperly formulated packages can cause.

The book does not, however, attempt to describe the chemistry of the additives or their manufacture in any great detail. The reason for omitting these aspects is that the range of formulations and chemical structures is too great for a book of this type and much of it is of a proprietary nature. The patent literature is full of additive formulations but most of them are not used commercially and so are of academic interest only.

A chapter on the use of oxygenates in gasoline has been included, even though oxygenates are not strictly additives but are more accurately regarded as blend components. It was considered that since they have been widely publicized as 'lead replacements', such a chapter was justified. Their use in diesel fuel, however, has been so limited that a corresponding section covering this aspect was not thought to be worthwhile.

The term 'gasoline' rather than 'petrol' has been used throughout since it was thought that it would be more familiar to most of the likely users. Only brief explanations of some of the terms relating to gasoline and diesel fuel technology have been given, and for more information the reader is referred to two general reviews on the subject [1,2].

Finally, I would like to express my warmest thanks to all the contributors for the care and effort they have put into each chapter. My thanks also to the numerous other people who have helped in so many ways in the production of this book and particularly to Lubrizol Limited and the Esso Research Centre at Abingdon.

K. Owen

References

1. E. G. Hancock, (ed.), *Technology of Gasoline. Critical Reports on Applied Chemistry*, Vol. 10, Blackwell Scientific Publications, London (1985).
2. G. D. Hobson (ed.), *Modern Petroleum Technology*, 5th ed., Part 2, Chapter 20, 'Fuels for Spark Ignition Engines' by K. Owen and Chapter 21, 'Diesel Fuels' by J. Boddy, John Wiley and Sons, Chichester (1984).

1 Precombustion gasoline additives

K. Owen
Middle Farm, Church Lane, Brightwell-cum-Sotwell,
Wallingford OX10 OSD
and
R. G. M. Landells
Lubrizol Limited, Waldron House, 57–63 Old Church Street,
London SW3 5BS

1.1	Introduction	2
1.2	**Additives providing oxidation stability**	**4**
1.2.1	Gasoline antioxidants	4
	(a) The mechanism of hydrocarbon oxidation	5
	(b) Storage stability	6
	(c) Lead stabilization	6
	(d) Engine fuel system cleanliness	7
	(e) Types of antioxidant	7
1.2.2	Metal deactivators (MDAs)	10
1.2.3	Oxidation stability test methods	10
1.2.4	Selection of optimum oxidation stability additives	11
1.3	**Additives used in gasoline distribution**	**12**
1.3.1	Dyes and markers	12
1.3.2	Corrosion inhibitors	12
1.3.3	Biocides	13
1.3.4	Anti-static additives	14
1.3.5	Drag-reducing agents	15
1.3.6	Demulsifiers and dehazers	15
1.3.7	Odorants	16
1.4	**Additives used to protect vehicle fuel systems**	**16**
1.4.1	Corrosion inhibitors	16
1.4.2	Demulsifiers and dehazers	17
1.4.3	Anti-icing additives	19
	Evaluation of anti-icing additives	21
1.4.4	Carburettor detergents	23
1.4.5	Fuel distribution improvement additives	26
1.4.6	Port fuel injector anti-fouling additives	27
1.4.7	Manifold, inlet port and inlet valve deposit control additives	29
1.4.8	Factory fill additives	34

| 1.5 | Detection of surfactant additives in gasoline | 35 |
| 1.6 | References | 36 |

1.1 Introduction

There are a large number of additives that can confer benefits before the fuel reaches the combustion chamber of an engine. These may be added to ensure that the gasoline does not deteriorate on storage, that it is free from problems during distribution and that deposit formation and corrosion do not occur in the fuel system of an engine. They may also be required for marketing or legal reasons.

The need for these additives has been growing for many years because of changes in gasoline composition arising from three main causes:

1. *Crude oil price changes.* Although crude oil prices have been rather volatile in recent years it is generally accepted that, in the longer term, because of the limited nature of crude oil reserves, they must increase very significantly. After the first oil crisis in 1973, prices rose rapidly for several years, and this had the effect of considerably reducing the use of fuel oil in power stations, steelworks, shipping, etc. in favour of other energy sources such as coal, hydroelectric and nuclear. The resulting excess of fuel oil left the oil companies with a severe imbalance in their product requirement [1], and there has been a massive move towards converting this heavy material to lighter products such as gasoline and middle distillates. This trend is expected to continue for several years to come.

    The conversion of heavier to lighter streams involves cracking the large hydrocarbon molecules and, unless accompanied by hydrogenation, it inevitably means that olefinic compounds are formed. Olefins are much more reactive than the other hydrocarbons present in gasoline and oxidize and polymerize readily to form gummy deposits. These deposits upset the metering of fuel and air into the combustion chamber and, because of this, can cause vehicle driveability malfunctions (i.e. hesitations and stumbles during accelerations and surging during cruise conditions). They can also result in sticking of valves and some other moving parts.

2. *Measures to improve exhaust gas emissions.* The need to reduce exhaust emissions of hydrocarbons, carbon monoxide and nitrogen oxides has been recognized for many years, and legislation to control these emissions has become increasingly restrictive and severe [2–4]. At

present, in many countries it is necessary to use exhaust catalyst systems in new car models to achieve the required levels, and this has meant that unleaded fuel is required. Concerns about the toxicity of lead itself have also played a large part in forcing the introduction of unleaded fuels.

Compared with leaded, unleaded gasoline needs higher octane quality components in order to achieve the specified octane levels, and this has given rise to major changes in composition. These, in turn, have increased the need for additives in many cases. The use of oxygenated components, for example, has meant that gasolines are more likely to cause corrosion problems and therefore have a greater need for corrosion inhibitors. Some oxygenates also increase the dissolved metals present in gasoline because they are corrosive to such metals as aluminium, magnesium, etc. Since these metals catalyse oxidation reactions there is a greater need for metal deactivators and antioxidants to counteract this problem. More severe catalytic reforming can give rise to heavy aromatic molecules which, together with any gums present, can lay down as deposits in the fuel intake system, needing detergents/dispersants to prevent excessive build-up.

The use of exhaust gas recirculation to reduce nitrogen oxide emissions can also worsen intake system deposits, since the exhaust gases contain particulates such as partially oxidized hydrocarbons which can deposit on surfaces in the intake system. PCV (positive crankcase ventilation) valves also operate in an environment that makes them susceptible to sticking unless suitable additives are present in the gasoline.

Port fuel injectors and more sophisticated carburettors, to provide better control of air/fuel ratios under all driving conditions, rely on absence of deposits to maintain their performance over the life of the vehicle. Again, detergent/dispersant additives in the fuel can play a large part in helping to achieve this end.

In the USA, vehicles must be capable of meeting the stipulated exhaust emission requirements without adjustment over a distance of 50 000 miles. Under these conditions it is extremely important that gasolines do not lay down deposits which will degrade the quality of the exhaust emissions. The use of appropriate gasoline additives helps to ensure that this need is met.

3. *Requirement to improve fuel consumption.* Fuel economy is an extremely important marketing aspect for motor vehicles and has been so ever since the price of fuel started to rise in 1973. The amount of fuel

consumption control required by governments varies throughout the world. In Europe there is no legislation for mandatory reductions in fuel consumption, although voluntary agreements have been made by motor manufacturers in several countries and there is a requirement that fuel consumption data be published. In the USA and Japan there is enforcement of staged improvements in fuel consumption [2].

The motor industry continues to make considerable progress in improving the fuel economy of vehicles. Two important steps that they have taken and which are influenced by additives are to maximize compression ratios as far as possible, consistent with fuel octane quality, and to use leaner mixture strengths.

High compression ratios improve the thermal efficiency of an engine but at the same time increase its octane requirement [5]. Changes in gasoline composition are necessary to achieve an acceptable octane level in the absence of lead, as has already been discussed, and this can require a greater use of additives. Leaner mixture strengths to improve fuel economy tend to make a vehicle more prone to driveability malfunctions and hence more critical to anything that leans the mixture strength still further, such as oxygenates. Again, additives have been used to reduce mixture maldistribution between cylinders (see Section 1.4.5), which is particularly important under lean mixture conditions. They are also used to ensure that deposits in critical parts of the system are not allowed to build up.

Thus it can be seen that precombustion gasoline additives can play an extremely important part in ensuring that vehicle performance is maintained as mileage is accumulated, in spite of pronounced changes in gasoline composition and vehicle design.

1.2 Additives providing oxidation stability

1.2.1 Gasoline antioxidants

Gasoline, in common with most organic materials, is subject to deterioration due to oxidation, and this occurs both during storage and in use in an engine. Oxidation gives rise to the formation of gums which can seriously influence the performance of the gasoline.

Gum formation first became a problem with the introduction of commercial cracking processes in the early 1920s, since the olefinic compounds produced by cracking are more susceptible to oxidation than the other hydrocarbons present. The use of these processes is still increasing quite rapidly due to the surplus of the heavy fractions from crude oil.

Of the olefins formed, the presence of diolefins is particularly undesirable, since these compounds are highly reactive and form gums and polymers very readily indeed. When gums are formed, they initially remain dissolved in the gasoline but, as the amount increases, they begin to separate out of solution. The problems that can then occur are blockages of lines and filters, high sludge levels in storage tanks, cloudy gasoline and deposits in various parts of an engine fuel system which can give rise to vehicle malfunctions.

Changes can be made in refinery processing to overcome these problems, such as the use of hydrogen treatment, but are usually relatively expensive since they often reduce yield and lower octane quality. Additives known as antioxidants or oxidation inhibitors retard oxidation, and these are now universally used as a cost effective alternative or supplement to processing modifications.

(a) The mechanism of hydrocarbon oxidation Gum is the product of a series of oxidation and polymerization reactions involving mainly the olefinic compounds present in a gasoline. The initial reaction product is a hydroperoxide followed by a series of other reactions giving rise to a complex range of products. The generally accepted mechanism can be illustrated as follows:

1. *Chain initiation*

$$R-H \rightarrow R\cdot$$

2. *Chain propagation*

$$R\cdot + O_2 \rightarrow R-O-O\cdot$$
$$R-O-O\cdot + R'H \rightarrow R-O-OH + R'\cdot$$

3. *Chain termination*

$$R\cdot + R\cdot \rightarrow R-R$$
$$R-O-O\cdot + R\cdot \rightarrow R-O-O-R$$
$$R-O-O\cdot + AH \rightarrow R-O-O-H + A\cdot$$

In the initiation stage, free radicals are generated by the homolytic cleavage of a hydrocarbon—this can occur either thermally or due to the presence of oxygen. Once a hydrocarbon free radical (R·) is formed it can consume a molecule of oxygen to form a peroxide radical (R—O—O·), which in turn can react with a further hydrocarbon molecule thereby generating another hydrocarbon free radical. The oxidation process is therefore self-perpetuating and is only terminated, in the absence of an antioxidant (AH), by reactions which lead to non-free-radical products. Peroxides themselves are pro-knocks and so are particularly undesirable in gasoline.

Free radicals can also give rise to polymerization as well as oxidation reactions to form high molecular weight materials that deposit in those parts of the intake system where the fuel is largely vaporized. Antioxidants function by combining with peroxide free radicals and by decomposing hydroperoxides into stable substances.

(b) Storage stability The primary purpose of an antioxidant is to extend the period for which a gasoline can be stored before its gum content becomes too high for trouble-free use. In time, all the antioxidant in a gasoline will be consumed and then gum formation will increase very rapidly. An antioxidant cannot destroy gum that has already been formed and, because of this, it is essential to add it as early in the refinery processing sequence as possible before the oxidation chain reactions have started. It is normal with cracked streams for the antioxidant to be injected in the rundown line from the process unit to tankage. Additional antioxidant can also be added to the finished gasoline blend.

The response of fuels to an antioxidant in terms of storage stability depends upon the fuel composition, and particularly upon the presence and type of olefinic compounds and the nature of the antioxidant used. Blend stability is not related proportionally to the stability of the individual components present, and the performance of an inhibitor in a single blending component is not a reliable criterion of its activity in a finished fuel or in a different component.

(c) Lead stabilization Antioxidants are not only valuable in preventing gum formation during gasoline storage but also in protecting lead antiknock compounds, particularly tetraethyl lead (TEL), from decomposition. Tetramethyl lead (TML) is much more thermally stable and so is less at risk.

TEL degradation usually manifests itself by a cloudiness in the gasoline, a light-coloured deposit settling out at the bottom of the tank and a loss of octane quality. The deposit will block filters, etc. in the automotive fuel system and can therefore cause driveability problems.

Usually only a low level of antioxidant is needed to protect lead compounds, and in most gasolines there is a considerable excess of additive to ensure adequate storage stability. Gasoline containing no cracked components may not need an antioxidant for storage stability reasons but it is important to ensure that enough is present for TEL stabilization—usually about 10 ppm* is adequate.

(d) Engine fuel system cleanliness In a vehicle the fuel may be stored for some time in the tank and other parts of the fuel system; when the vehicle is operating the gasoline is vaporized and mixed with cold or hot air before entering the combustion chamber. It is subject, therefore, to ideal oxidation conditions before being burnt, although, fortunately, the time for such reactions to occur is usually rather limited. The antioxidants employed to improve the storage stability of gasoline are most effective when used in a liquid fuel since they do not vaporize very readily. In those parts of the fuel system of a vehicle where the fuel is completely or partially vaporized (such as in the inlet manifold, around the ports, on the undersides of inlet valves, etc.) protection is very much diminished, and so the effectiveness of an oxidation inhibitor is rather restricted. However, some protection against deposit formation can be given by other types of additive (see Section 1.4).

The nature of the deposits (i.e. whether they are soft and sticky, or hard), the position in the intake system where they deposit and the rate of formation will depend upon a number of factors. These include the degree of polymerization/oxidation that has occurred, the temperature involved and the presence of additives in the fuel. Particulates and other products from combustion can be pulsed back into the fuel inlet system because of valve overlap. These materials attach themselves to the sticky deposits and increase the volume and modify the nature of the deposits.

(e) Types of antioxidant [6] The effectiveness of an oxidation inhibitor depends upon its chemical structure, the composition of the gasoline to be protected and the conditions under which the gasoline is to be stored. It is important for the antioxidant to be readily soluble in gasoline at all temperatures and to be as insoluble as possible in any aqueous layer that may be present in the bottom of a gasoline storage tank. It should not react with any of the fuel components or other additives, and it should be capable of complete combustion without leaving any residual deposit in the combustion chamber.

* ppm refers to parts per million by weight (i.e. mg/kg).

Gasoline antioxidants in current use belong to two main classes of chemical compounds, i.e. aromatic diamines and alkyl phenols. To a very much smaller extent, amino phenols are also used, but these compounds, although possessing high activity, suffer from the disadvantage of having a relatively high solubility in water and caustic solutions and so can be quickly lost to water bottoms on storage. They have also been reported to contribute to inlet system deposits.

(1) *Aromatic diamines.* *Para*phenylenediamines are extremely effective antioxidants, particularly for gasolines having a high olefin content, and are usually used in the range of 5–20 ppm. A number of compounds of this type are used commercially and have the general formula:

$$R-NH-C_6H_4-NH-R'$$

The groups R and R' can be the same or different and are commonly sec-butyl, isopropyl, 1,4-dimethylpentyl or 1-methylheptyl. One of this class of antioxidant that is frequently used is NN'-di-sec-butyl-p-phenylenediamine. It has a freezing point of about 14°C and, although it often supercools, it is best used with a diluent at low ambient temperatures.

Although water bottoms in gasoline storage tanks are often somewhat alkaline, it is not impossible for them to be acidic. This can happen when certain processes are used with an inadequate caustic wash and when insufficient time has been allowed for water from such processing to settle out in rundown tankage. In these circumstances there is a risk that some of the diamine antioxidant will be extracted into the water layer and thereby reduce the protection of the gasoline against oxidation.

Other potential problems associated with the amine type of antioxidant are:

1. Discoloration of light-coloured paintwork on vehicles if there is a spillage of fuel down the side of the car during filling, due to oxidation of the molecule.
2. Coloration of the gasoline and/or the water bottoms due to reactions with other compounds present in the gasoline.

On the other hand, this type of antioxidant can generally be used at very much lower concentrations than the alkyl phenol type, and it is often used only for cracked stocks and where the total olefin content of the gasoline is high.

Aromatic diamines have also been used for 'inhibitor sweetening', in which the additive acts as a catalyst in the oxidation of the evil-smelling mercaptans to the non-odorous disulphides using dissolved oxygen. The process is rather slow and is no longer widely used except for the removal of traces of any mercaptans left after Merox or similar refinery treatments.

(2) *Alkylphenols*. This type of antioxidant is most effective in gasolines having a low level of olefins, i.e. below about 10 per cent by volume. The most important phenols having good antioxidant activity in gasoline have sterically hindered hydroxyl groups due to the presence of alkyl groups in the 2, 6 positions. The most commonly used are probably 4-methyl-2, 6-di*tert*-butylphenol, 2, 4-dimethyl-6-*tert*-butylphenol and 2, 6-di*tert*-butylphenol. These can be supplied either alone or mixed with other alkyl phenols. The mixtures have the advantages that they have lower freezing points than the pure materials and can be more cost effective.

Steric hindrance of the hydroxyl group reduces the solubility of the molecule in alkaline water. The water bottoms in refinery gasoline storage tanks are frequently somewhat alkaline. Alkylphenol antioxidants generally suffer fewer adverse side effects than the diamine type but need to be used at higher concentrations to get the same effectiveness— usually 5–100 ppm.

(3) *Mixtures of aromatic diamines and alkylphenols*. These are reported to outperform equivalent concentrations of either constituent alone. The ratio of diamine to phenolic type used in a mixture depends upon the

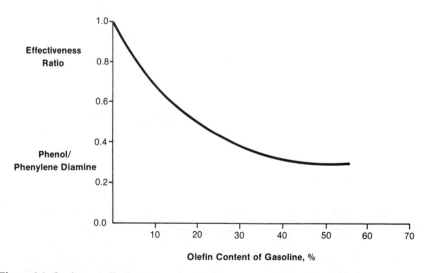

Figure 1.1 Optimum effectiveness ratio of phenolic to amine antioxidant versus gasoline olefin content.

olefin content of the gasoline. The higher the olefin content, the higher the relative amount of diamine type to be used, as illustrated in Figure 1.1 [7].

1.2.2 Metal deactivators (MDAs)

Trace levels of soluble metal compounds, particularly copper, catalyse the oxidation of hydrocarbons and can lead to the very rapid formation of high levels of gum. Metal deactivators overcome this problem by chelating onto the metal, rendering it inactive.

It is often considered that metal deactivators are only necessary when copper-sweetening processes are used in refineries. In fact, the majority of trace metals present in gasoline come from the reaction of acidic compounds present (e.g. mercaptans, phenols, etc.) with copper and various alloys used in refinery equipment, distribution systems and vehicle fuel systems. When some alcohols are incorporated into a gasoline blend a whole range of other metals can be attacked and dissolve in it. Many of these metals are potential catalysts for oxidation and polymerization reactions. The most widely used metal deactivator is N, N'-disalicylidene-1, 2-propanediamine, which forms a compound with copper having the structure shown in Figure 1.2. For optimum effectiveness, MDAs should be added prior to tankage but downstream of a caustic wash unit, since they can be extracted by a caustic solution. The addition rate is normally in the range of 4–12 ppm. They are usually sold as a solution to prevent them freezing in cold weather.

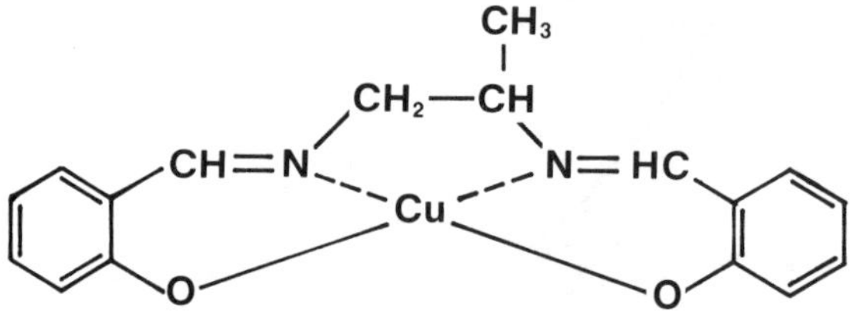

Figure 1.2 Copper chelate.

1.2.3 Oxidation stability test methods

The most common test procedures [8,9] used to establish the gum content of gasoline and its storage stability are as follows:

1. *Induction Period* (ASTM D525/IP 40). In this test a sample of the gasoline is introduced into a bomb and subjected to a high pressure of oxygen at 100°C. The pressure in the bomb is continuously recorded

and the time when a sharp pressure drop occurs is noted. This time indicates that oxygen is starting to be consumed very rapidly and that gum is being formed. It has to be said, however, that although the induction period is widely used as an indication of the storage stability of a gasoline the correlation with gum-forming tendency is poor and will depend upon storage conditions and gasoline composition. Under average European storage conditions an induction period of 240 min corresponds to a storage life of about two years.
2. *Existent gum* (ASTM D381/IP 131). This test measures the amount of gum that is actually present in the gasoline at the time of testing and gives no indication of the likelihood of a gasoline to form gums. In the method a measured quantity of fuel is evaporated in a glass beaker under controlled conditions and the residue is weighed before and after extraction with n-heptane. The heptane-insoluble part is known as the Existent Gum, and the evaporation residue prior to heptane treating is the Unwashed Gum. A maximum Existent Gum level of 4 mg/100 ml is usually considered satisfactory. The Unwashed Gum figure indicates the total amount of non-volatile materials present, but, since it includes most of the additives, it does not give a good guide to gasoline performance.
3. *Potential Gum* (ASTM D873/IP 138). This test is claimed to measure how much gum it is possible to form in a gasoline. It is carried out by conducting an Induction Period test for a fixed time (often 16 h) and then carrying out an Existent Gum test on the oxidized gasoline.
4. *Other tests.* A number of other tests are used by various companies to assess the oxidation stability of gasoline blends, and include the Copper Beaker Test and the Three Months' Dark Storage Test. In the Copper Beaker Test 50 ml of gasoline is evaporated in a polished copper beaker under standard conditions and the residue (the Copper Beaker Gum) is weighed. The Dark Storage test is carried out at 43 °C and at the end of 3 months an Existent Gum test is performed. This latter test gives a good indication of storage stability but is of little value for routine control use.

1.2.4 Selection of optimum oxidation stability additives

It is important when deciding on which additive to use that any test work carried out is with the actual gasoline and not with a simulated blend made up in a laboratory from components taken from tankage. This is because if there are any cracked materials present they should already have been treated with antioxidant in the rundown line from the cracking unit. The recommended steps to take are:

1. Measure how satisfactorily the current additive performs.
2. Discuss with additive suppliers alternatives that are likely to be more cost effective than the present material, including the use of metal deactivators.
3. Carry out a *plant* trial by substituting the new additive for the existing one. Monitor the performance carefully over a period of at least 2 weeks before coming to any decision.

1.3 Additives used in gasoline distribution

1.3.1 Dyes and markers

The most common reasons for using dyes are first, to meet legal requirements, as in the USA, where all leaded gasoline must be coloured, and second, to be able to distinguish one product or brand from another. They are also used, with or without special marker chemicals, to supply evidence in cases of theft, tax evasion, fuel adulteration, etc. and for identifying the source of leaks.

The dyes are usually red, orange, blue or green, and are mainly azo compounds although, in the case of the blue dye, this is often an anthraquinone. Treat levels are very low and normally in the range of 2–10 ppm, since it is important that gasolines containing them do not stain the sides of light-coloured vehicles if there is a spill during filling. Dyes supplied as concentrated solutions are preferred by refiners because they are easier to handle.

Marker chemicals are particularly important for security purposes since they impart no colour to the fuel but are detected either by a colour reaction with another chemical or by other means. Such markers must be detectable without interference by lead alkyls, dyes and other gasoline additives. Furfural or diphenylamine are often used as markers and are detected by mixing the gasoline with another chemical to give a colour. The presence of marked fuels can generally be detected in concentrations as low as 5 per cent in other fuels.

1.3.2 Corrosion inhibitors (see also Section 1.4.1)

Internal corrosion of pipelines is a serious problem, since it reduces flow rates and can give rise to suspended rust in the gasoline. The main purpose of rust inhibitors is to minimize corrosion in pipelines, although the remainder of the fuel system and storage tanks also benefit. For this purpose, these additives are used at comparatively low treat rates—typically, below 20 ppm.

The additives themselves are surfactant materials that attach themselves as a monomolecular layer to the internal surface of the pipeline, thereby protecting it from attack. Many different chemical types are used but all have a polar or hydrophilic group at one end of the molecule and an oleophilic/hydrophobic group such as a long chain alkyl group at the other. The polar group attaches itself to the metal surface and the non-polar tail sticks into the hydrocarbon phase and provides an oily layer that repels water.

The most effective polar groups are acidic (e.g. carboxylic, phosphoric and sulphonic acids). The amine salts of these acids are also effective. The use of additives containing phosphorus has, however, now almost disappeared because of their adverse effect on exhaust gas catalysts.

In recent years the widespread use of alcohols as gasoline octane boosters has increased the corrosivity of gasoline, and this has raised the need for distribution system corrosion inhibitors. Usually, higher concentrations of additive are required to prevent corrosion when alcohols are present, not only because of the increase in dissolved water but also because the solvent action and polarity of the hydrocarbon/alcohol blend reduces the ability of a surfactant to attach itself to the pipeline walls. The procedures for measuring the effectiveness of anti-corrosion additives are discussed in Section 1.4.1.

1.3.3 Biocides

One of the beneficial side effects of the use of lead alkyls in gasoline is their biocidal properties, i.e. they prevent microbial growth. However, the ever-increasing usage of unleaded gasoline could mean that this problem will become very much more common—as it is with middle distillates—so that the use of biocides in gasoline could grow very significantly.

When microbial activity commences in a gasoline tank the colour and clarity of the gasoline and the tank water bottoms change. The colour becomes somewhat darker and suspended matter develops at the fuel/water interface which can block filters, etc. Once it has started, it can be quite difficult to stop, but there are a number of actions that should be taken in addition to considering the use of a biocide [10]. These can include the following:

1. The fuel should be made less susceptible to microbial activity by avoiding the use of additives containing nitrogen or phosphorus;
2. The pH of the water bottoms in tankage should be adjusted to minimize the solubility of any gasoline additives present that contain nitrogen or phosphorus;
3. Free water should be removed by regular and frequent draining of water bottoms from storage tanks. It is in this free water that microorganisms proliferate and from which they obtain some of their nutrients.

A number of commercial biocides are available which have a wide range of chemical compositions (they may be boron compounds, imines, amines, imidazolines, etc.). The dose rate used is low but the treatment cost can be quite high. When using them it should be remembered that the dead organisms do not simply disappear but that they can be released from the sides of tanks and lines so that there may be an increase in suspended matter on first using the additive. It is advisable to carry out manual cleaning and desludging before using a biocide.

1.3.4 Anti-static additives

When fuels of low electrical conductivity are pumped through pipes, etc., particularly at high velocities, a charge of static electricity can build up in the fuel. This can give rise to sparking as it discharges, and if this happens within a flammable mixture of hydrocarbon vapour and air a fire or explosion can take place. Fortunately, when filling a vehicle with gasoline the vapour/air mixture coming from the tank is usually much too rich to burn or explode, and, in addition, the conductivity of gasoline is often high enough to dissipate the charge.

This problem of static electricity is particularly acute with aviation kerosine, which has a very low conductivity and is handled at high pumping rates. Anti-static additives have been added to this material for many years in order to increase its conductivity. The additive used has been mainly a chromium-based organometallic material, although totally organic additives are now available.

Concern is now being expressed over the possibility of static discharge problems with gasoline, since in some cases high pumping rates are being used. Some conventional surfactant additives are effective in reducing the conductivity, as are many oxygenates.

Precombustion gasoline additives 15

1.3.5 Drag-reducing agents (DRAs)

When pipelines become limiting in capacity one possibility to increase the throughput is to use drag-reducing additives in the products going through the pipeline. These additives are extremely high molecular weight polymers which shear very readily and reduce drag by smoothing turbulent bursts off the walls of the pipeline and so allow more product to flow. They have been used in some countries at concentrations of up to about 50 ppm. There are concerns about their long-term use and possible effect on fuel intake system deposits such as on valves and injectors, but it is probable that a properly formulated additive package containing appropriate detergents/dispersants will overcome any adverse influences.

1.3.6 Demulsifiers and dehazers (see also Section 1.4.2)

Water can find its way into fuel during refinery processing and distribution and can be present both as free and as dissolved water. Dissolved water can be forced out of solution by a sudden drop in temperature and will appear as a haze that can often take a long time to clear if the gasoline is simply left to stand on its own. Water can also come out of solution during blending if there is a temperature difference between components or if two saturated components having different water solubilities are blended together. Aromatic components tend to be able to dissolve more water than paraffinic ones.

The ease with which the water haze clears depends mainly upon whether emulsion-stabilizing materials are present such as some surfactant additives, finely divided solids, etc. Special anti-haze additives are available which are extremely effective in accelerating the rate at which the gasoline clears by promoting coalescence of small droplets. They are themselves surface-active materials which have limited solubility in both gasoline and water, and so tend to concentrate at the fuel/water interface.

When free water is present in the tank it can sometimes be entrained with gasoline during pumping and the shearing forces involved can give rise to quite stable emulsions that, in severe cases, gives the gasoline a milky appearance and can plug filters. This will normally only happen when excessive amounts of surfactant additives are used or when such additives have been improperly formulated. Anti-haze type additives can also help to avoid these problems, and are generally used in the range of 1–30 ppm.

Chemically, the anti-haze additives used in gasoline can be very complex, and often consist of a mixture of several different compounds. The composition is almost always proprietary. The selection of the most suitable additive and concentration to use is a matter of considerable skill and experience, since their effectiveness depends upon what other additives are present, the composition of the gasoline itself (including the use of oxygenates) and on the storage conditions. Because the anti-haze additives are themselves surfactants they can interfere with the effectiveness of other surfactants present in terms of detergency, etc. For this reason, if such additives are frequently found to be necessary to ensure consistently clear and bright gasolines it is advisable to check that the benefits of any other surfactants being used are still valid. Test procedures for quantifying haze and emulsion-forming tendency are given in Section 1.4.2.

1.3.7 Odorants

Some gasoline blend components have an extremely unpleasant odour due to the presence of certain compounds such as mercaptans. The almost universal use of self-service pumps at service stations has made it important for gasolines to have an odour that customers find acceptable.

It is preferable to overcome odour problems by refining processes such as hydrogen treatment, but the use of low concentrations of compounds such as vanillin (4-hydroxy-3-methoxy-benzaldehyde) has been tried on a limited scale. Unfortunately, some customers seemed to prefer the original gasoline odour to the modified one, so the optimum solution has probably yet to be found!

1.4. Additives used to protect vehicle fuel systems

1.4.1 Corrosion inhibitors (see also Section 1.3.2)

Corrosion in the fuel system of a vehicle can lead to severe problems. Not only can leaks develop in automobile fuel tanks but particles of rust can block fuel lines, filters and critical carburettor orifices such as jets. Small amounts of water and dissolved air promote corrosion of ferrous parts of the fuel system. In addition, alcohols such as methanol or ethanol, present as blend components, can attack many non-ferrous parts, as well as increase the dissolved water content of the gasoline.

The types of compounds used to prevent corrosion in the fuel systems of vehicles and the way in which they function are the same as those used to protect distribution systems (see Section 1.3.2).

Two types of test are used to evaluate gasoline corrosion inhibitors:

1. Dynamic, to simulate transport by pipeline, use in vehicles, etc.
2. Static, to simulate storage conditions

The most common of the procedures used are [7–9]:

1. *Dynamic corrosion test* (ASTM D665/IP 135). This test was originally designed to evaluate the rust-preventing characteristics of steam-turbine oils and has been modified for application to gasoline. In the method a mixture of 300 ml of the gasoline under test is stirred with 30 ml water at a temperature between ambient and 38°C. A polished cylindrical steel specimen is completely immersed in the liquid and the test is usually run for 24 h. At the end of this time the spindle is rated for degree of rusting. The water used is often distilled to simulate rain water, but this should be supplemented with additional tests in which water at different pH levels (in the range 4–9) is used in order to simulate the types of process water that can find their way into gasoline.

   A number of other dynamic tests have been accepted for use, such as MIL-I-25017B, MIL-I-25017C, Colonial Pipeline Test and the National Association of Corrosion Engineers (NACE) Test.
2. *Static corrosion test.* For this procedure a 3-weeks test is often used in which 10 ml of the desired type of water and 90 ml of gasoline are put into a stoppered bottle and a strip of carbon steel is then added. After shaking, the bottle is left undisturbed for the required test period with estimates of the degree of rusting being made at suitable time intervals.

1.4.2 Demulsifiers and dehazers (see also Section 1.3.6)

The use of additives to prevent haze and emulsion formation has already been discussed in Section 1.3.6, with special reference to distribution system problems. It would be unusual for these water-sensitivity problems to commence after the fuel has reached the vehicle's tank.

Many surfactant additives are extremely effective at emulsifying water in gasoline and at keeping small particles in suspension. For this reason, most commercial surfactant additive packages designed for induction system deposit control, anti-icing, corrosion inhibition, etc. have incorporated into them a small amount of anti-haze additive to minimize these problems. It would be undesirable, therefore, to consider adding more of these materials to a gasoline containing such an additive package, since by doing so one could upset the delicate balance between different surfactants and possibly destroy the benefits that should accrue. When selecting an additive for use it is worthwhile to review carefully the data supplied by additive manufacturers showing the water sensitivity and particulate suspension characteristics of their surfactant additive packages to ensure that adequate protection is given.

The most common performance evaluation tests used [7–9] are given below. With all of them it is worthwhile to run extra tests on additive-free gasoline in order to be able to evaluate the true effect of the additive.

1. *Water tolerance* (ASTM D1094/IP 289). A mixture of 80 ml of gasoline is shaken with 20 ml of the desired type of water for 2 min and the amount of haze in the gasoline layer and of emulsion at the interface is rated after 5 min settling period. It is worth carrying out this test using a number of different pH waters (in the range 4–9) in order to simulate better the field situation.
2. *Multiple-contact emulsion test.* This test gives very much more information than the simple ASTM D1094 test on how an additive-treated gasoline will perform in practice, since it attempts to simulate what happens in a gasoline tank where the water bottoms stay relatively constant and successive batches of gasoline are passed through. The test requires 10 ml water and 90 ml gasoline in a stoppered 100 ml measuring cylinder to be shaken and then allowed to stand for 24 h. The appearance of both the gasoline and the interface is then rated using a special scale. The gasoline layer is next carefully poured away, leaving the water behind, and is then replaced with a further 90 ml of the test gasoline. The mixture is once again shaken, allowed to stand for 24 h, and the gasoline and interface rated as before. This process is repeated for a total of ten or more cycles. For gasolines with inadequate water shedding properties, haziness can appear in the gasoline due to dispersed water and emulsions can form at the water/fuel interface after a number of cycles.
3. *Other tests for water sensitivity.* There are a large number of 'in-house' tests used which include rapid and severe stirring of the gasoline with water in a blending machine such as the Waring Blender.
4. *Rouge suspension test.* This test establishes the likelihood of a gasoline to hold particulates in suspension. It involves shaking 0.1 g of precipitated iron oxide (rouge) in 100 ml gasoline and measuring the amount remaining in suspension after given time intervals.

1.4.3 Anti-icing additives

Intake system icing is a phenomenon associated mainly with carburetted cars, although ice formation in throttle-body-injected vehicles has also been known to occur. It should not be confused with the freezing of extraneous water in fuel lines, etc. It is noticeable during idle and/or cruise conditions and results in loss of engine power, high fuel consumption, increased CO and hydrocarbons in exhaust gases and engine stalling. Figure 1.3 shows the range of ambient conditions that can give rise to it.

Icing is caused by moist intake air condensing and freezing in the carburettor or throttle body due to the evaporation of the gasoline. It only occurs in cool humid weather, i.e. when the ambient temperature is between about -5 and $12\,°C$ and the relative humidity is above about 80 per cent. At lower temperatures there is not enough water vapour present in the air to cause any severe problems, and at higher ones the temperature depression caused by the vaporizing gasoline is not sufficient to allow ice to be formed. Icing is worsened by increasing gasoline volatility and is strongly dependent on engine design. Vehicles are tending to become less susceptible to this problem due to improvements in carburettor design and in inlet air temperature control. However, quite a surprising number of cars still suffer from it in geographical areas where cool, humid weather persists.

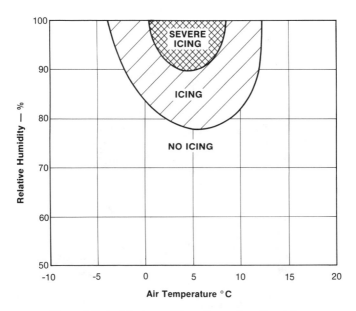

Figure 1.3 Ambient conditions for carburettor icing.

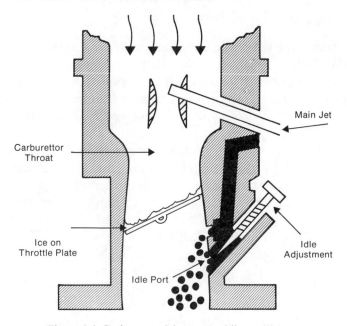

Figure 1.4 Carburettor icing under idle conditions.

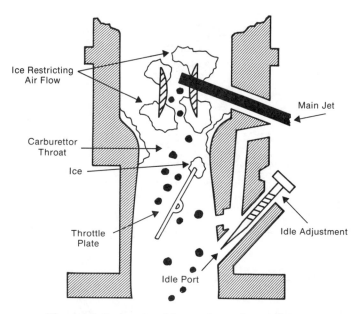

Figure 1.5 Carburettor icing under cruise conditions.

When ice forms on the throttle plate it causes a richening of the air/fuel mixture at very low throttle openings such as when the vehicle is idling, and this causes stalling (see Figure 1.4). This type of icing is known as idle icing. Ice formation in the venturi area (see Figure 1.5) also restricts air flow and causes power loss and stalling under cruise conditions (cruise icing).

Two additive approaches for controlling icing are available:

1. Freeze point depressants (cryoscopic additives), and
2. Surfactants

The freeze point depressants act in the way the name suggests, and are mainly low molecular weight polar compounds such as alcohols and glycols that are soluble in water. Isopropyl alcohol and dipropylene glycol are two specific materials that have been used for this purpose, as also are mixtures of different glycols and other water-soluble oxygenated compounds. These materials vary widely in their effectiveness and 2 per cent of isopropyl alcohol will give about the same benefit as 0.1 per cent of dipropylene glycol. Where alcohols such as methanol, ethanol, etc. are used as blend components in gasoline they are extremely effective as freeze point depressants, and no further anti-icing additive addition is necessary in such cases.

The surface-active anti-icers function by the same mechanism as corrosion inhibitors, i.e. by preventing water from reaching the metal surfaces and hence interfering with the adhesion of the ice crystals to the throttle plate and other parts. A general correlation exists between anti-rust effectiveness and anti-icing performance. A wide range of chemical types are used as anti-icing additives, and these include imidazolines and carboxylic acid salts.

Evaluation of anti-icing additives Various test procedures have been developed to demonstrate the icing characteristics of gasolines and measure the effectiveness of additives in reducing intake system icing. The only satisfactory and credible tests involve using an engine or a car, and the main difficulty is that it is necessary to control both the temperature and the humidity of the inlet air to the engine. The other problem is that cars and engines vary widely not only in the extent to which they suffer from icing but also in the type of icing that occurs. If a given engine is selected as the basis for a test it may respond differently towards additives than many other vehicles in the car population under consideration.

Most additive manufacturers use 'in-house' engine tests in which the time for sufficient ice to build up to cause stalling is employed as a method of comparing the effectiveness of different additives. This is quite satisfactory in most cases, except that it covers only one type of icing in one particular engine.

Because of the unsatisfactory nature of the existing tests, the British Technical Council (Driveability Group), which consists of representatives from the oil, motor and additive industries, have developed some new test procedures which were then taken up on a European basis by the Coordinating European Council (CEC) [11]. Three such test methods were developed and all of them involve running a car on the road or, preferably, on a chassis dynamometer in a chamber where the ambient temperature and relative humidity are controlled. They cover urban driving icing, cruise icing and idle icing, and, in summary are carried out as follows:

1. *Urban driving icing.* With this procedure the tendency for the engine to stall during idle in city driving is evaluated. This type of icing occurs mainly while the engine is warming up and before inlet air heating devices are operating. Stalling can be a nuisance, particulary with automatic transmission vehicles.

    In the test, a car critical to this type of icing is driven using a fixed cycle, preferably on a chassis dynamometer under controlled conditions of temperature and humidity (usually something like 5 °C and 95 per cent relative humidity, since these represent severe icing conditions). The cycle involves a full-throttle acceleration to 50 kph, then a cruise at 50 kph for 30 sec when the vehicle is halted and allowed to idle for 30 sec. The vehicle must not have been warmed up prior to the start of the test, and ten of these cycles are run. The whole procedure is intended to represent city driving with fast accelerations between traffic lights. During the idle mode on each cycle the quality of the idle and the occurrence of stalls is recorded and an overall numerical rating assigned.

2. *Cruise icing.* This test is mainly concerned with ice formation in the venturi area of the carburettor, which gives rise to loss of power and, in extreme cases, stalling of the engine while cruising at a steady and moderate speed. It can occur when the vehicle is fully warmed up and is often very puzzling to the motorist since, when the car has been halted for a few minutes, heat soak-back from the engine melts the ice and the vehicle behaves perfectly normally again. It is exacerbated by the fact that in some vehicles inlet air heating does not take place during cruise

conditions when the vehicle is fully warmed up, for fuel economy reasons.

The procedure involves running a test car, known to be susceptible to this type of icing, at a constant *throttle* position to give a selected speed, usually in the range 70–120 kph, under ambient conditions critical to icing. As ice forms in the carburettor the vehicle loses speed and may eventually stall. The test lasts for 30 min and is normally carried out with a fully warmed-up vehicle. When evaluating additives the difference is used between the area under the plot of vehicle speed versus time when no icing occurs (obtained by testing with up to 5 per cent isopropyl alcohol or other cryoscopic additive in the base gasoline) and the area when running with the additive in the gasoline.

3. *Extended idle test.* This test was developed at the request of several motor manufacturers, who indicated that some of their customers like to start their cars and allow them to idle unattended in the mornings while they finish breakfast, so that they have a warm car ready to drive away. Under these conditions stalling due to ice formation can occur if the weather conditions are appropriate. The procedure enables motor manufacturers to modify their inlet air heating systems to minimize this problem, and is less often used to evaluate additives. It is carried out by allowing the vehicle to idle under controlled conditions and the number of stalls in a given time is used as a measure of the degree of icing.

In all these tests it is important to use a relatively high-volatility gasoline as the base test fuel so as to ensure that readily perceived differences are obtained.

1.4.4 Carburettor detergents

The carburettor is an intricate metering device designed to mix fuel and air in the correct proportion for each engine-operating condition. Deposits can accumulate on the walls of the throttle body, on the throttle plate, in the idle air circuit and in metering orifices/jets as shown in Figure 1.6. These deposits can upset the fuel/air ratio and cause driveability malfunctions. When they form in the throttle plate area they cause a restriction in the air supply when the throttle is at or near the closed position, and this gives rise to rough idling, stalling, increased fuel consumption and increased exhaust emissions. Deposits in the jets can affect all aspects of driveability performance adversely.

The principal sources of contaminants contributing to deposits are

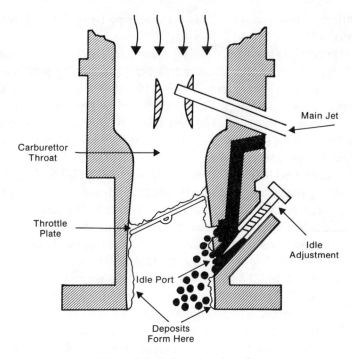

Figure 1.6 Carburettor deposits.

exhaust fumes which can be pulsed back into the intake system due to valve overlap or taken in through the air inlet; blowby gases; and, when the air cleaning system is faulty, dust. It is, however, the oxidation stability of the gasoline itself which mainly controls the accumulation of deposits. Unsaturated hydrocarbons (olefins and diolefins) in the fuel oxidize to form gums and resins which, in turn, act as binders for the other contaminants. Some of this oxidation takes place in the carburettor when the engine is switched off and heat soaks back into this area. For this reason, stop/start type driving is often the most critical in terms of carburettor deposit formation. Other sticky oxidized materials come from the blowby and exhaust gases, and these also play a part in the initial laydown of deposits and the binding together of particulates, etc.

Carburettor detergents can prevent these deposits from forming and can also clean up those that have already been laid down. The mechanism by which they function has not been well defined but it is generally agreed that the polar groups in the detergent preferentially attach themselves to metal surfaces and to any particulates present since these also have a polar nature. The non-polar part of the molecule sticks out into the fuel hydrocarbons so that a monomolecular film is formed on the surfaces, preventing deposition and particle aggregation. Particles are thus carried on into the combustion chamber and burnt. Deposits that have already been laid down can gradually be removed by this mechanism, which in effect solubilizes them. Usually a higher concentration of additive is needed for clean-up than for keep-clean effectiveness.

Conventional amine detergent additives effectively prevent the build-up of deposits when used at 20–60 ppm and will clean up pre-formed deposits at higher concentrations. However, as treatment levels approach and exceed 100 ppm this type of additive may begin to increase induction system deposits, depending on the temperature regime in the engine [12] (see also Sections 1.4.6 and 1.4.7).

Amine detergent additives can be categorized chemically as amines, alkanol-amines, amides, amino-amides and imidazolines (see Figure 1.7). Although the primary function of these chemical types is carburettor detergency, they can also impart some anti-corrosion and anti-icing properties to the gasoline, and in this regard are multifunctional. The multifunctionality of amine detergents can be enhanced to a significant

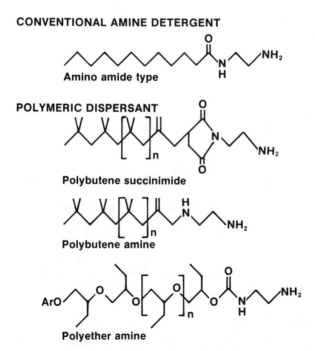

Figure 1.7 Detergent and polymeric dispersant gasoline additives.

degree by neutralizing either partially or fully with carboxy or orthophosphoric acids (but see Section 1.3.2 regarding the declining use of phosphorus-containing additives).

A large number of engine test procedures have been used to measure the effectiveness of carburettor detergents, and most of them are in-house methods that have been developed by additive manufacturers and oil companies for their own use. There are, however, some procedures which have been cooperatively developed for general use, i.e.:

1. The Ford Carburettor Cleanliness Test, developed by CRC in the USA
2. The Renault 5 Carburettor Cleanliness Test, developed by CEC in Europe [13]
3. The Twin Carburettor Opel Kadett Engine Test, also developed by CEC in Europe [14]

The first two of these tests are summarized in Table 1.1 and the Opel Kadett test, which is more often used to evaluate intake valve deposits, in Table 1.2 (see Page 33).

Table 1.1 US and European carburettor detergency tests

Test method	CRC Ford Carburettor Cleanliness Test	CEC Renault 5 Carburettor Test (CEC F-03-T-81)
Engine type	Ford 3.9-litre Six-cylinder	Type 810–26 four-cylinder
Test duration	20 h	12 h
Cycle	3 min at 700 rpm 7 min at 2000 rpm	2 min at 800 rpm 8 min at 1800 rpm (4.3 KW)
Cooling water temperature	88–90 °C	80 °C
Engine oil temperature	110 °C max.	78 °C
Intake air temperature	66 °C	40–65 °C
Exhaust gas recirculation	None at idle Full at cruise	11 per cent
Reference oil	REO-202-TI SAE 30 Grade	CEC RL-51 10W50 Grade Dispersant PMA

1.4.5 Fuel distribution improvement additives

Fuel maldistribution between the cylinders of cars designed for lean carburation can result in increased exhaust emissions, driveability

Precombustion gasoline additives

problems and some loss of fuel economy. The inlet manifold usually has a film of liquid gasoline on its walls. This incompletely evaporated fuel may find its way into the cylinders in an uneven way, giving poor cylinder-to-cylinder mixture distribution. Oleophobic additives, such as hydrogenated tallow amines (HTA) [15], form a coating of low surface energy on the walls of the induction system which inhibits the spread of gasoline as a thin film, so that it forms, instead, a number of discrete droplets. These droplets are more readily entrained by the air stream and so give improved fuel distribution.

1.4.6 Port fuel injector anti-fouling additives

Increasingly stringent fuel economy and emission control requirements have forced vehicle manufacturers to move towards injectors for more precise fuel metering. These inject the fuel into the intake manifold or the intake port rather than directly into the cylinders. They can be electronically or mechanically operated.

Port fuel injection (PFI) is, at present, the best means to meet both fuel economy and exhaust emission limits whilst maintaining or improving overall performance. The number of PFI-equipped cars is therefore increasing considerably. The pintle type gasoline fuel injector is currently the most popular design, and a sectional view is shown in Figure 1.8.

Fuel metering occurs in the annular region between the pintle tip and the surrounding outer shell. These injectors are very sensitive to small levels of deposits in this metering zone (Figure 1.9) and such deposits can

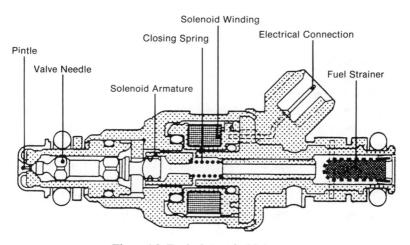

Figure 1.8 Typical port fuel injector.

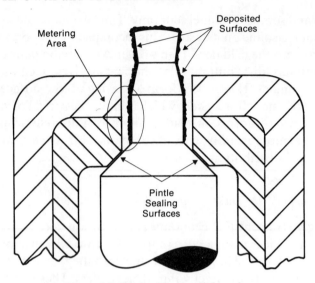

Figure 1.9 Pintle tip area of deposits.

reduce fuel flow and alter the injector spray pattern. This can make cars difficult to start and can cause severe driveability malfunctions, loss of power, an increase in exhaust pollutants and a decrease in fuel economy. Such fouling problems reached major proportions in the USA in 1986, but are now claimed to have been largely overcome by the use of additives and by the introduction of less critical injectors [16–20]. Although a closed-loop fuel metering control system provides some compensation for the restriction in flow caused by deposits it cannot compensate for different levels of restriction among injectors as a result of non-uniform deposit formation.

Unlike carburettor deposits, PFI deposits are created when the fuel is not flowing through the injectors. The rate of deposit formation is primarily affected by the hydrocarbon composition of the base gasoline [19] and, in particular, the presence of olefinic components increases the deposit-forming tendency. It is believed that small amounts of gasoline in the annular region evaporate during heat soakback when the engine is switched off, since temperatures in the region of 90–100 °C can occur at the injector tip. The heavier aromatic and olefinic gasoline molecules left behind gradually oxidize and polymerize to form gums and resins. These adhesive materials can then hold particulate material, formed during combustion, in place.

PFI deposits are formed at much higher temperatures than carburettor deposits and consequently are much more difficult to deal with. The same general types of conventional amine detergent additives that are effective in controlling carburettor deposits have proved to have some effectiveness for port fuel injectors, but higher concentrations are necessary. However, as discussed previously, high levels of these conventional amines (more than 100 ppm) can lead to an increase in inlet manifold and inlet valve deposits.

To overcome this problem, polymeric dispersant systems (i.e. dispersants and fluidizers together) with better thermal stabilities have been used to control PFI deposits. While amine detergents are limited to carburettor/PFI deposit control and rust control in the fuel system, these polymeric dispersants were found to control deposits throughout the induction system (see Section 1.4.7).

Technically, there is a functional overlap between detergents and dispersants. The difference is primarily the action by which cleanliness is accomplished. Detergents emphasize a cleansing action, while dispersants include cleansing but additionally have the ability to disperse particulate matter in an extremely fine and harmless state and hold it in suspension. In this way it will pass through the fuel system into the combustion chamber and burn with the fuel.

Testing for additive effectiveness in terms of overcoming PFI fouling is usually carried out on a chassis dynamometer using a vehicle known to be critical to this problem. One method developed and reported in the USA [7, 12, 18] uses a light duty cycle involving 15 min driving at 55 mph (88 kph) and road load followed by 45 min with the engine off to allow heat soakback into the fuel system. This one-hour cycle is repeated until driveability malfunctions manifest themselves or until an average injector plugging of 10 per cent is reached. Plugging of injectors can be quantified by removal from the engine and measuring liquid flow rates through them.

PFI deposit formation in this test correlates well with field experience. A laboratory procedure to simulate PFI deposit build-up is currently under development in the USA under the auspices of the Coordinating Research Council (CRC). Although in Europe and other countries the problem has not been so widespread as in the USA, some vehicles have been reported as suffering from it and a similar chassis dynamometer procedure has been used to investigate it.

1.4.7 *Manifold, inlet port and inlet valve deposit control additives*

An increasingly severe problem in recent years has been the formation of induction system deposits in the manifold, inlet ports and on the inlet

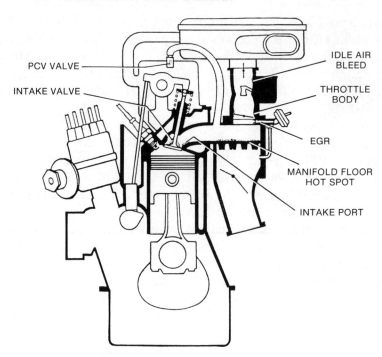

Figure 1.10 Induction system deposits (carburettor engine).

valves (see Figures 1.10 and 1.11). These deposits can cause driveability problems such as rough idle, stalling and poor acceleration as well as increased exhaust pollutants and fuel consumption. In the manifold, deposits in the 'hot spot' area, used to help vaporize the gasoline, can be particularly important, since they can increase warm-up times and influence driveability.

The deposits can vary in both structure and consistency, depending upon the composition of the gasoline, whether gasoline additives are present, the type of additive, the temperatures in the induction system and the general design of the system. They may be soft and sticky, greasy, firm or hard with open porous or closed surfaces. Porous deposits can act as a sponge and absorb fuel, making the air/fuel ratio too lean during the initial warm-up period and thus lead to poor engine performance. They may build up sufficiently to interfere with the closing of the valve, and this can cause valve burning, poor breathing and loss of power and fuel economy. Tacky deposits can sometimes find their way up the valve stems so that when the engine is stopped and allowed to stand under cold ambient conditions these deposits harden and cause valve sticking. In

Precombustion gasoline additives

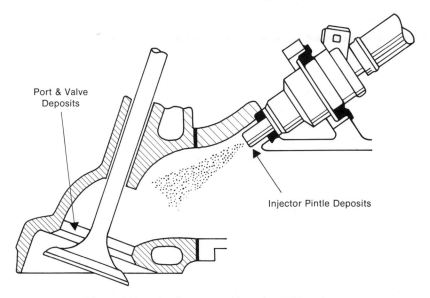

Figure 1.11 Induction system deposits (PFI engine).

severe cases, hard valve deposits may break off, pass into the combustion chamber and become lodged in the ring belt or wedged between the piston crown and cylinder head or valves.

Vehicles vary widely in their propensity towards induction system deposit formation, and so engine design is a major factor in determining whether high or low levels of deposit will form. The adoption of positive crankcase ventilation (PCV) and the use of exhaust gas recirculation (EGR) for nitrogen oxides control have aggravated the deposit problem. The other two important contributors to these deposits are the gasoline and the lubricating oil. Higher inlet valve temperatures and an increase in valve diameter in some modern European lean-burn engines have also worsened the situation. Some such engines require only very low levels of valve deposits before driveability malfunctions become apparent.

Thermally and oxidatively unstable gasoline components, such as those containing olefins and diolefins, are often the precursors of valve deposits, as is the case in other parts of the fuel system. The use of alcohols in some areas has worsened this deposit problem and may also be a contributory factor to black sludge formation in the crankcase and other parts of the engine. Additive chemistry has been shown to influence quite strongly the formation of black sludge in the lubricant [21].

The lubricating oil that passes between the valve guide and the valve stem is also known to contribute to valve deposits. Lubricating oil, unless it is fresh, contains oxidized materials and contaminants, which, on further exposure to air and high temperatures, can deposit high molecular weight materials on the inlet valves. The contaminants can be derived from the gasoline by crankcase dilution during cold starts, and it is suspected that complex reactions sometimes take place with lube oil additives to give sticky, black, sludge-like materials that deposit on valves and which also can harden in the valve guides in cold weather and cause valve sticking. Certain viscosity index improvers present in multigrade oils have also been known to increase inlet valve deposits.

Polymeric dispersants are effective in controlling induction system deposits and in helping to maintain vehicle performance, fuel economy and exhaust emissions to levels close to those provided by a new engine. These additives can also improve cleanliness in the crankcase by sludge and varnish control. It is customary to combine such additives with petroleum-based or thermally stable synthetic oils known as fluidizers, carrier oils or solvent oils. When these high boiling oily materials reach the valve tulips they seem to dissolve the sticky, deposit-forming compounds, which are then swept on into the combustion chamber. They also reduce the tendency for them to coke on the hot metal surface. Care, however, must be taken in formulating, since an improperly balanced polymeric dispersant/fluidizer package may give rise to valve sticking.

Fluidizers alone can be quite effective, even without the use of polymeric dispersants, in controlling inlet valve deposits, and such materials as heavy lube oil base stocks (e.g. 500 Solvent Neutral base oil or certain polybutenes) have been used for this purpose. Continuous treatment at high dosages can, however, result in high levels of combustion chamber deposits, increased octane requirements of the engine [7] and thickening of the lubricating oil.

Although the increase in octane requirement of an engine with mileage is considered to be mainly due to combustion chamber deposits (see Chapter 2) it has been reported [22] that ridge-like deposits on the intake port can have as great an effect on octane requirement as combustion chamber ones. Such deposits can be controlled by polymeric dispersants.

Chemically, polymeric dispersants are relatively high molecular weight materials and can be divided into three chemical groups—succinimides, polybutene amines and polyether amines (Figure 1.7). They are added to fuel at higher concentrations (20–600 ppm) than conventional amine detergents. The addition of fluidizer, corrosion inhibitor, demulsifier, exhaust valve seat protection additive (see Chapter 2) and solvent can

Precombustion gasoline additives 33

result in a multifunctional gasoline additive package with a total treat rate of 1000–2000 ppm.

Polyether amines have been reported [23] to give good induction system cleanliness, including inlet valves, without the use of a fluidizer. They also have the advantage of not increasing the viscosity of the lube oil when crankcase dilution occurs. The mechanism by which these polymeric dispersants function has been discussed in Section 1.4.6.

Several engine test procedures exist for evaluating the effect of additives on induction system deposits. In Europe a procedure using an Opel Kadett engine was developed in the late 1970s by DKA in West Germany and was later refined by a CEC group [14] (Table 1.2). Although the test is superior to laboratory procedures such as gum or oxidation stability tests in predicting the performance of a gasoline to form (or otherwise) intake system deposits, the engine no longer represents current technology. It does not, therefore, give results that are necessarily applicable

Table 1.2 Opel Kadett inlet system deposit test CEC F-04-A-87

Test conditions			
40 h keep-clean test	Time (min)	Speed (rpm)	Power (kW)
Cycle (4.5 min)	0.5	950	—
	1.0	3000	11.1
	1.0	1300	4.0
	2.0	1850	6.3
Coolant outlet temperature 92 °C max.			
Oil temperature 94 °C max.			
Intake air heating to PCV 86–90 °C			
Reference oil CEC RL-51 10W50 grade dispersant PMA VII			
Engine Opel Kadett 12S			
Four-cylinder, four-cycle 1.2-litre twin carburettor gasoline engine			

Table 1.3 M 102E inlet system deposit test

DB test conditions				
150 h keep-clean test				
Cycle	Time	Speed	Load	Power
(4.5 min)	(min)	(rpm)	(newtons)	(kW)
	0.5	800	—	—
	1.0	1300	31	4
	2.0	1850	34	6.3
	1.0	3000	37	11
Temperatures				
Cooling water outlet	90–95 °C			
Engine oil	95–100 °C			
Intake air	25–35 °C			

to vehicles of more recent design and particularly to those using fuel injectors instead of carburettors for mixture control.

Daimler Benz have developed a new procedure involving the use of a 2.3-litre M 102E injection engine. This procedure satisfies many current requirements and is being used by many oil companies and additive manufacturers in Western Europe. The Daimler Benz operating conditions and temperatures are given in Table 1.3 [24] but, at the time of writing, there are many variations of the test in use and it has not yet been accepted as a CEC procedure. The most common test durations used are 60 or 150 h without oil change.

Other modern European engines currently used in 'in-house' testing of induction system deposits include the 1.05-litre VW Polo, the 2.5-litre BMW 525 and the Porsche 944 engine. A 1.9-litre water-cooled Boxer engine has also been used to study valve sticking [26]. In the USA induction system deposit control in modern lean-burn engines is being studied to a large extent using imported European and Japanese vehicles [25, 27].

No matter how successful an additive package is in controlling induction system deposits using a bench engine or a chassis dynamometer test procedure, the ultimate testing of any additive system is always an extensive field test. Although this is time consuming and costly, it is the only credible way to demonstrate the effectiveness of additives in practice and to identify the possibility of problems occurring.

There is clearly a need to standardize some of the performance tests for the evaluation of additives for overcoming induction system deposits, to ensure a common understanding of their effectiveness.

1.4.8 Factory fill additives

In times of industrial recession when sales have been depressed it has been necessary for many car manufacturers to stockpile vehicles for quite long periods. These will have been driven to the storage area and so will have a certain amount of gasoline in their tanks and fuel system. Because of this, it is common practice for many Original Equipment Manufacturers (OEMs) to use gasolines containing a special blend of additives to minimize the possibility of corrosion and oxidation stability problems occurring. Several OEMs have a list of approved factory fill additives.

Such additive packages will involve the use of unusually high-treat levels of antioxidants, metal deactivators, corrosion inhibitors, biocides, etc. The precise formulation will vary according to the nature of the fuel system in which it is to be used, i.e. the metals present, the nature of any

Precombustion gasoline additives 35

plastic or elastomeric materials, etc. Another problem suffered by OEMs that can be overcome by factory fill additives is that of plug fouling, which arises because of the frequent need to shuttle new cars for short distances prior to final delivery to the customer. These runs are almost always made with full use of choke, so that carbon fouling of the plugs eventually occurs, which prevents restarting. This can be overcome by the use of selected mineral oils, polymers or dispersant formulations in the gasoline. Factory fill fuel additive packages are also used by various defence forces for military vehicles in storage.

1.5 Detection of surfactant additives in gasoline

It is sometimes necessary or desirable to be able to analyse a gasoline to see what type and concentration of surfactant additive is present. This may be in order to check on the accuracy of the injection equipment used or to establish the losses of additive that might be occurring during distribution due to adsorption on to rust, losses to tank water bottoms, etc.

It must be said that it is extremely difficult to carry out such analyses, particularly if one is attempting to identify the nature and concentration of an unknown additive. Additives are commonly ashless, and today, because of concern for exhaust catalyst poisoning, most commercial additives contain only carbon and hydrogen together with other elements such as sulphur, nitrogen and oxygen. Distinguishing these compounds at low concentrations from other materials present in gasoline presents a challenge.

There are no commonly available test procedures for this purpose, although it is usual to evaporate off the bulk of the gasoline at a low temperature and reduced pressure and then to investigate the residue by extraction with different solvents followed by the use of such techniques as infrared spectroscopy, liquid chromatography, etc.

Additive manufacturers will often be able to supply procedures for identifying and quantifying their own additives in gasoline, and these will normally involve thin layer chromatography (TLC) and high-performance liquid chomatography (HPLC). In these cases, where the neat additive is available it is possible to make up standard strength solutions of the additive in gasoline to compare with the results obtained on the test sample.

It must be remembered that, in many cases, additive packages can contain several different surfactants. These can be lost in the distribution system at different rates, depending upon their individual polarities, water solubilities, etc. It can be important to quantify all the components present to be sure that the gasoline has all the claimed benefits. In some cases it may be easier to simply test for performance effectiveness—for example, by carrying out anti-corrosion laboratory tests, etc. rather than detailed chemical analyses.

1.6 References

1. C. W. C. Van Paassen, 'Changing refining practice to meet gasoline and diesel demand and specifications requirement', *Proceedings of the Institution of Mechanical Engineers* (November 1986), Paper C316/86.
2. CONCAWE, *Report No. 4/88*, J. M. Tims *et al.*, 'Trends in motor vehicle emissions and fuel consumption regulations' (April 1988).
3. H. R. Glatz, 'The historic development, the political background and the future perspectives of motor vehicle emission control and emission control regulation in Europe', *Institution of Mechanical Engineers Symposium* (1987), Paper C358/87.
4. V. Korte and D. Gruden, 'Possible spark-ignition engine technologies for European exhaust emission legislations', *Institution of Mechanical Engineers Symposium* (1987), Paper C334/87.
5. G. D. Hobson (ed.), *Modern Petroleum Technology*, 5th ed., Part 2, Chapter 20, 'Fuels for Spark Ignition Engines' by K. Owen, John Wiley and Sons, Chichester (1984).
6. J. H. Gilks, 'Antioxidants for petroleum products', *J. Inst. Pet.*, **50**, No. 491, 309–17 (1964).
7. R. C. Tupa and C. J. Dorer, 'Gasoline and diesel fuel additives for performance/distribution quality—2', Society of Automotive Engineers, Warrendale, SAE Paper No. 861179 (1986).
8. American Society for Testing and Materials, *Annual Book of ASTM Standards*; Section 5; Petroleum Products, Lubricants and Fossil Fuels; Vol. 05.01–05.05. American Society for Testing and Materials, Philadelphia (1989).
9. Institute of petroleum, *Methods for Analysis and Testing of Petroleum and Related products*, Vols 1 and 2, John Wiley and Sons, Chichester (1989).
10. E. C. Hill, *Treatment Strategies*, Presented at Institute of Petroleum Microbiology Committee Conference on 29 October 1986, pp. 55–61.
11. Coordinating European Council, Tentative Test Method No. CEC M-10-T-87, 'Intake system icing procedures for use on road, track or vehicle dynamometer with spark ignition vehicles'.
12. R. C. Tupa and D. E. Koehler, 'Gasoline port fuel injectors—keep clean/clean up with additives', Society of Automotive Engineers, Warrendale, SAE Paper No. 861536 (1986).
13. Coordinating European Council, Tentative Test Method No. CEC F-03-T-81, 'Evaluation of gasolines with respect to maintenance of carburettor cleanliness'.

14. Coordinating European Council, Test Method No. CEC F-04-A-87 (Was Tentative Test Method No. CEC F-02-T-79), 'The evaluation of gasoline engine intake system deposition'.
15. A. A. Zimmerman, L. E. Furlong and H. F. Shannon, 'Improved fuel distribution—a new role for gasoline additives', Society of Automotive Engineers, Warrendale, SAE Paper No. 720082 (1972).
16. D. L. Lenane and T. P. Stocky, 'Gasoline additives solve injector deposit problems', Society of Automotive Engineers, Warrendale, SAE Paper No. 861537 (1986).
17. B. Y. Taniguchi, R. J. Peyla, et. al., 'Injector deposits—the tip of the intake system deposit problems', Society of Automotive Engineers, Warrendale, SAE Paper No. 861534 (1986).
18. G. P. Abramo, A. M. Horowitz and J. C. Trewella, 'Port fuel injector cleanliness studies', Society of Automotive Engineers, Warrendale, SAE Paper No. 861535 (1986).
19. J. D. Benson and P. A. Yaccarino, 'The effects of fuel composition and additives on multiport fuel injector deposits', Society of Automotive Engineers, Warrendale, SAE Paper No. 861533 (1986).
20. R. C. Tupa 'Port fuel injector deposits—causes/consequences/cures', Society of Automotive Engineers, Warrendale, SAE Paper No. 872113 (1987).
21. H. P. Rath, K. Starke and H.-H. Vogel, Wirkung optimierter Kraftstoffadditive in ausgewählten Prüfmotoren, Technische Arbeitstagung, Hohenheim (1988).
22. L. B. Graiff, 'Some new aspects of deposit effects on engine octane requirement increase and fuel economy', Society of Automotive Engineers, Warrendale, SAE Paper No. 790938 (1979).
23. R. A. Lewis, H. K. Newhall, et al., 'A new concept in engine deposit control additives for unleaded gasolines', Society of Automotive Engineers, Warrendale, SAE Paper No. 830938 (1983).
24. M. Gairing, 'Zur Qualität der Ottokraftstoffe aus der Sicht der Automobilindustrie; Vermeidung von Ablagerungen auf Einlassventilen', Mineralölrundschau, 34, No. 11, 209–15, November (1986).
25. B. Bitting, F. Gschwendtner, et al., 'Intake valve deposits—fuel detergency requirements revisited', Society of Automotive Engineers, Warrendale, SAE Paper No. 872117 (1987).
26. S. Mikkonen, R. Karlsson and J. Kivi, 'Intake valve sticking in some Carburetor engines', Society of Automotive Engineers, Warrendale, SAE Paper No. 881643 (1988).
27. R. C. Tupa and D. E. Koehler, 'Intake valve deposits—effects of engines, fuels and additives', Society of Automotive Engineers, Warrendale, SAE Paper 881645 (1988).

2 Gasoline additives influencing combustion processes

D. R. Blackmore
Shell Research Ltd, Thornton Research Centre,
PO Box 1, Chester CH1 3SH

2.1	Introduction	40
2.2	Anti-knock additives	43
2.2.1	Lead anti-knock additives	44
2.2.2	Other metallic anti-knock additives	48
2.2.3	Co-anti-knock additives	50
2.2.4	Ashless anti-knock additives	50
2.3	Anti-ORI (octane requirement increase) additives	52
2.3.1	Halogen scavenger additives	53
2.3.2	Boron deposit modifier additives	53
2.3.3	Detergent additives used intermittently at a high dose rate	54
2.3.4	A detergent additive used continuously at a low dose rate	54
2.4	Anti-run-on additives	55
2.5	Anti-pre-ignition additives	55
2.6	Anti-misfire additives (or driveability additives)	56
2.6.1	Spark plug anti-foulant additives	56
2.6.2	Spark-aider additives	57
2.6.3	Additives for improving fuel distribution	58
2.7	Anti-VSR (valve seat recession) additives	59
2.7.1	Lead additives	59
2.7.2	Phosphorus additives	59
2.7.3	Alkaline metal additives	59
2.7.4	Other materials	60
2.8	Additives that improve lubricant performance	60
2.8.1	Upper-cylinder lubricants/friction modifiers	60
2.8.2	Anti-wear additives	61
2.8.3	Anti-sludge additives	61
2.9	Miscellaneous additives	62
2.10	Future prospects	62
2.11	References	63

2.1 Introduction

The combustion processes that take place in the spark-ignition gasoline engine are far from ideal, and in considering the role of additives in assisting combustion we need to appreciate the designer's ultimate purpose as well as his more practical targets.

To obtain the maximum thermal efficiency from the burning of the hydrocarbon fuel it would ideally be necessary to release the heat energy of the fuel under constant-volume conditions. Such behaviour would require that the combustion take place instantaneously and homogeneously, with no variation from one engine cycle to the next, nor from one year to the next. Such ideal combustion behaviour of the fuel and air mixture would in turn require that the spark-ignition event would be perfectly repeatable, that the flame process would be both infinitely fast and repeatably so and that there would be no heat losses to the walls of the combustion chamber and cylinder and no emission of partially burnt or other undesirable combustion products.

In practice, this essentially thermodynamic picture of the spark-ignition engine is, of course, subject to the limitations of both chemical kinetics and fluid dynamics of the fuel and air as they are fed through the reciprocating engine. Thus the fuel and air do not react instantaneously; instead they combust together in what is designed to be a fully pre-mixed turbulent flame that travels as quickly as possible but in reality only relatively slowly across the combustion chamber. The flame is initiated by a spark event whose timing is arranged to be somewhat ahead of the movement of maximum compression so as to allow for the slowness of the subsequent flame. The art and science of engine timing over the years has been to so arrange the spark timing and the mixture preparation that under all possible engine conditions, transient as well as steady running, this 'normal' state of engine combustion will always prevail.

However, in practice a number of 'abnormal' combustion characteristics can intrude (Figure 2.1):

1. Knocking combustion (also known colloquially as pinking) occurs when the gas mixture most remote from the flame initiation auto-ignites before the flame can reach it. This end-gas explosion causes a rapid rise in local pressure, and this expanding gas meets the advancing turbulent flame with its expanding gas behind it, with the result that very high oscillatory pressures are set up. The effect is easily audible, but can also be picked up by accelerometers as vibrations are set up in the whole engine structure. High temperatures are rapidly reached as

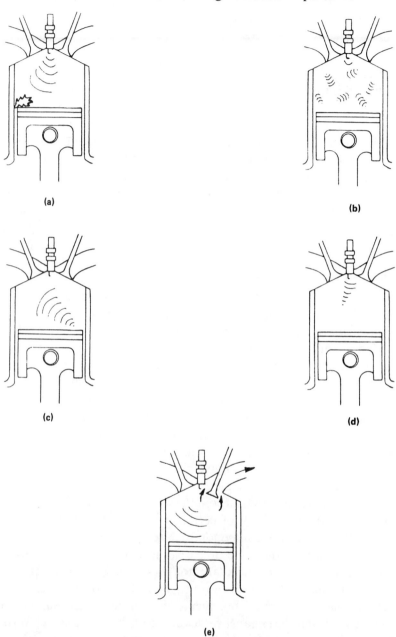

Figure 2.1 (a) Knock (an end-gas explosion); (b) run-on (a compression ignition at idle); (c) pre-ignition (a surface ignition before spark-ignition can occur); (d) misfire (flame kernel too weak and so the flame dies early); (e) exhaust valve seat recession (VSR) (environment too hostile for the 'soft' metallurgy).

heat transfer is increased many times, and engine damage can soon take place.
2. Run-on combustion occurs when the engine continues to fire even after the electrical spark-ignition is switched off. The fuel–air mixture is inducted into the hot combustion chamber by the momentum of the engine, and because of high enough heat transfer processes, the mixture can auto-ignite (or 'diesel') for a short time until the engine cools.
3. Pre-ignition can take place in more than one way, thereby establishing a flame before the main spark ignition event has occurred. One common source is the hot spark plug itself, thus locating the pre-ignited flame in the same place as usual. Another is the surface of the piston, whereby hot metal or a hot layer of piston deposit can also pre-ignite the incoming gases. The problem with pre-ignition is that very often it can become a runaway process, leading to catastrophic overheat and engine damage.
4. Misfire, totally or partially, is a relatively common abnormality that is more of an inconvenience than a source of engine damage. It occurs, for example, when the fuel–air mixture is poorly mixed (giving rise to fuel maldistribution among the cylinders) or is not fuel-rich enough (giving rise to a lean-limit extinction of the flame), or the electrical spark is unable to deliver enough energy (as a consequence of deposits or fouling on the spark plugs). In each case the newly formed flame kernel is not allowed to grow and develop into a self-sustaining flame. Instead, the flame merely dies out after only a few millimetres of travel. The engine is felt to stumble or hesitate or give other symptoms of poor driveability. In the extreme, bad misfire leads to considerable power loss and nowadays can also give rise to exhaust catalyst burnout.
5. Exhaust-valve seating damage can occur under certain circumstances, and while this is not strictly a combustion abnormality in itself it can be attributed to the activity of certain combustion products. Exhaust valve seat wear is a phenomenon that occurs if too high temperatures are reached and at the same time certain 'solid-lubricant' combustion products (e.g. lead products) are not present in significant quantities. The phenomenon has been resolved by use of resistant steels for the exhaust valves and seats, but recurred when lead was removed from gasoline without the exhaust valves and seats being made of improved, resistant metallurgy.

Each of the above areas of combustion abnormality can be tackled by engine-design approaches and also by the adoption of modifications to the fuel by means of additives. In some cases the first route overtakes the need for the second, in others, vice versa. In what follows, we treat each of the areas in turn from the gasoline additive standpoint. We deal with each case both historically and with an eye to the present and near-future: this is particularly significant at a time when European gasoline is going through its biggest change for several decades by virtue of the introduction of unleaded gasoline for the first time into many markets.

2.2 Anti-knock additives

Very early in the evolution of the spark-ignition engine the phenomenon of knock was encountered. However, it was not until the early systematic work in 1910–30 of Ricardo [1] in the UK and of Kettering and others [2–4] in the USA that the phenomenon became well enough understood for real inroads to be made into its solution from the viewpoint of fuel design.

The phenomenon needed to be properly defined and measured, and in the years between World Wars I and II the engine-testing procedures were set in place with the adoption of the CFR (Cooperative Fuels Research) single-cylinder test engine. At this time the role of hydrocarbon composition came to be recognized and the octane scale was invented. The relationship of fuel behaviour in the CFR engine to that in a vehicle on the road was also explored. A fuller account of these events can be read in Robinson's recent article [5].

For our present purposes it is sufficient for us to note that the mechanism of knock can be attributed to an end-gas explosion or auto-ignition before the normal flame can reach it. Therefore for fuel design or additives to affect the situation they must act either to increase the speed of the flame or to decrease the propensity of the end-gas mixture to auto-ignite. Since the flame is a turbulent flame process owing much of its speed to the physical processes of heat transfer from the burnt to the unburnt gases rather than to any chemical ones, the likelihood of finding chemical additives that increase this flame speed is rather remote. However, the process of auto-ignition is different, it being mechanistically a free-radical chemical chain reaction whose rate can be readily controlled by appropriate chemical radical inhibitors or traps.

It is well known that lead alkyls have provided the most cost-effective anti-knock additives for the last 60 years. However, it is less well known that other metallic-based materials have also been used (and in some cases still are). Much early research was undertaken on organic materials (e.g.

amines), and more recently searches for ashless anti-knocks have again been attempted.

2.2.1 Lead antiknock additives

The serendipitous discovery of lead alkyls as anti-knocks, their use in the automotive market which led to the creation of a new lead-additive industry and, currently, their impending eclipse in the face of environmental pressures make a graphic and fascinating story of our times (see also Robinson [5]). We will restrict ourselves to a more detailed technical approach.

Lead alkyls have been found to be highly effective anti-knock additives on a weight (and cost) basis. As little as 0.1%w gives a very substantial benefit in octane number to a base fuel. A useful survey has recently been provided by Russell [6]. Several suitable lead alkyls exist: the four alkyl groups that combine with lead must be light in order for the additives to remain sufficiently stable and volatile, and hence tetraethyl lead (TEL) and (since 1960) tetramethyl lead (TML) have been the principal materials used. Chemical mixtures of these additives can also be used (i.e. methyltriethyl lead, dimethyldiethyl lead and trimethylethyl lead). Table 2.1 lists the properties of these materials [7], and Figure 2.2 shows a typical improvement in octane number as a function of lead content, commonly referred to as a lead susceptibility plot [8].

However, lead alkyls cannot be used on their own. Because of problems arising from the deposition of refractory lead oxide materials in the combustion chambers, which in the early days caused the failure of spark plugs and exhaust valves, scavenger materials were needed. Halogenated ethanes here proved to be very effective cures of these problems. Either 1, 2-dibromoethane or the cheaper 1, 2-dichloroethane can be used in quantities theoretically equivalent to the amount of lead present (giving rise to the concentration units of 'Theories').

In practice, several variations in the lead anti-knock additive package are possible (see Table 2.2). Typically, Aviation Mix is used for aviation gasolines and consists usually of TEL-B, tetraethyl lead with 1.0 Theory of 1, 2-dibromoethane. Motor Mix is commonly used for motor gasolines and consists usually of PM50-CB, i.e. a physical mixture of 50/50 TEL/TML together with 1.0 Theory of 1, 2-dichloroethane and 0.5 Theory of 1, 2-dibromoethane.

As lead levels have been reduced following environmental legislation in many countries a reappraisal of the effectiveness of lead alkyls has been

Table 2.1 Properties of pure lead alkyls

Chemical formula	Tetraethyl lead $(C_2H_5)_4Pb$	Methyltriethyl lead $(CH_3)(C_2H_5)_3Pb$	Dimethyldiethyl lead $(CH_3)_2(C_2H_5)_2Pb$	Trimethylethyl lead $(CH_3)_3(C_2H_5)Pb$	Tetramethyl lead $(CH_3)_4Pb$
Molecular weight	323.5	309.4	295.4	281.4	267.3
Pb content (% w/w)	64.06	66.96	70.14	73.64	77.50
Specific gravity 20°/4°C	1.650	1.714	1.790	1.882	1.995
Boiling point at 1.013 bar (°C)	200 (decomposes)	170	155	135	110
Vapour pressure at 20 °C (mbar)	0.34	1.00	2.9	9.7	31.5

Table 2.2 Compositions of lead anti-knock compounds (% w/w)

	TEL-B ('Aviation Mix')	TEL-CB	TML-CB	PM50-CB ('Motor mix')
TEL	61.5	61.5	0	30.8
TML	0	0	50.8	25.4
1,2-dibromoethane	35.7 (1.0 Theory)	17.9 (0.5 Theory)	17.9 (0.5 Theory)	17.9 (0.5 Theory)
1,2-dichloroethane	0	18.8 (1.0 Theory)	18.8 (1.0 Theory)	18.8 (1.0 Theory)
Dye and inerts	2.8	1.8	0.1	0.9

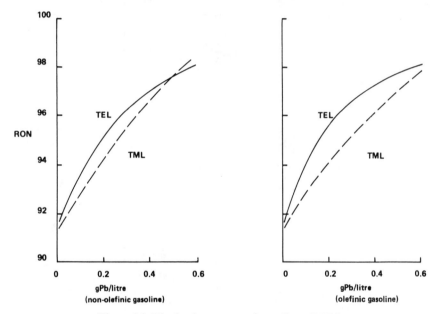

Figure 2.2 The lead response of gasolines (RON).

made for the lower lead concentrations in the gasoline produced by modern refinery streams. Figures 2.2–2.4 show the lead susceptibility plots [8] for TEL and TML on RON, MON and R 100 °C* in non-olefinic (cat-reformed) and olefinic (cat-cracked) types of gasoline. Examination of

* Research Octane Number (RON) is a measure of low-speed octane quality; Motor Octane Number (MON), of high-speed octane quality; and RON of the volatile fraction up to 100 °C in a standard distillation (R100 °C) is a measure of the octane quality under low-speed accelerating conditions.

Gasoline additives influencing combustion processes 47

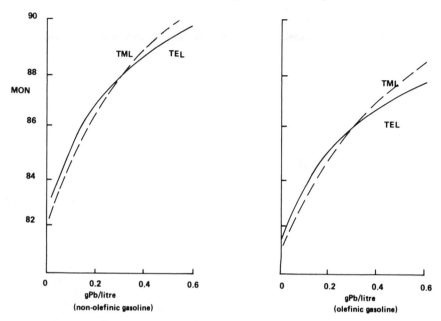

Figure 2.3 The lead response of gasolines (MON).

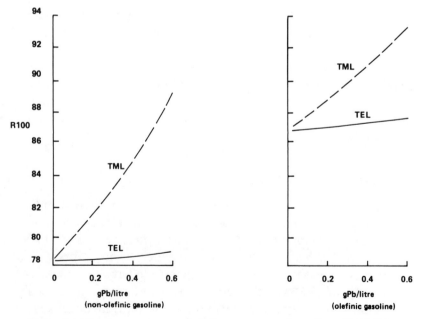

Figure 2.4 The front-end octane number response of gasolines (R100 °C).

these figures shows that, at low lead levels, TEL is most effective for RON, TEL and TML are about equivalent for MON and TML (because of its volatility) is much more effective for R 100°C. This type of information can be interpreted to mean in practice that there is real scope for mixed lead alkyl compounds (physical or chemical mixes) as follows:

1. For Regular (lower octane) gasolines, TEL is normally the most effective compound.
2. For Premium (high octane) gasolines, if RON is the most critical property, TEL is normally the most effective compound (but physical or chemical mixes are preferred at higher lead levels.
3. For Premium gasolines, if MON is the most critical property, physical or chemical mixes are preferred at all lead levels.
4. For Premium gasolines, if R100 °C is the most critical property, TML or physical mixtures are the most effective materials.

The future use of lead alkyls is clearly dependent on the advance of environmental legislation, although it is fair to point out that, in world market terms, lead will continue to be used in substantial quantities for many years yet. The reasons for its reduction are twofold: a concern over the toxicity of inorganic lead in the exhaust, for which a reduction is the environmentally cost-effective step: and a concern over its action to de-activate exhaust catalysts which are in use to control other undesirable exhaust pollutants, in which case its total removal is necessary from those gasolines used in such catalyst cars.

There are other factors in the debate over lead phase-out. One reason to retain at least some lead is its ability to lubricate exhaust valve seats, thus preventing seat recession (valve sinkage). Another is the belief that octane requirement increase is greater when lead is excluded. We return to both these points later.

2.2.2 Other metallic anti-knock additives

Over the years there have been many attempts to discover alternative organometallic anti-knock compounds. Very early in the 1920s selenium compounds were found [9] to be moderately effective, and this then led to the discovery of lead. Since then, manganese, iron, copper, thallium, cerium, nickel and tin have all been shown [6] to have had some beneficial effect (see Figure 2.5), and the very full patent literature has added others (cobalt, molybdenum, vanadium, titanium, lithium and boron). However, they all have had drawbacks, usually in terms of cost-effectiveness

Gasoline additives influencing combustion processes

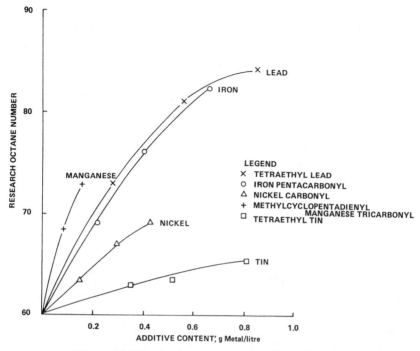

Figure 2.5 Effectiveness of organometallic anti-knocks.

(thallium, cerium, nickle and tin) or in terms of engine durability (iron, nickel and tin).

The most interesting and indeed best candidate has been manganese. The molecule that was found [10] to be suitable to carry the manganese is methylcyclopentadienyl manganese tricarbonyl (MMT), also known sometimes by its trade name AK-33X. This proved to be nearly as effective on a weight basis as lead (more so in paraffinic gasolines, less in aromatics). It also was found to be synergistic when used in conjunction with lead, and so was used commercially as a supplemental anti-knock from 1958. However, it has been only used commercially as a primary anti-knock in unleaded gasoline in the USA since 1974 at up to 40 parts per million by weight (ppmw) of manganese. It is also used currently in Canada, at a very low dose rate of about 20 ppmw manganese in unleaded gasoline.

However, even this very promising material fell in the USA at the hurdle of the effect on exhaust emissions. In a long and protracted technical debate in the late 1970s the additive manufacturer [11] failed to convince the EPA [12] and the US motor companies [13] that manganese was not harming the exhaust emissions performance, particularly of catalyst cars.

Work by the manufacture at low dose rates down to 5 ppw Mn (equivalent to an octane number boost of only 0.7) provided a possible comprise, but the motor companies were still unhappy if the EPA insisted that MMT be added to the certification fuel for the evaluation of catalyst durability.

MMT is thus left as a useful supplemental anti-knock for leaded fuels, but unless a supplemental additive or scavenger can be found to mitigate its effect on HC-producing engine deposits its widespread future as an anti-knock for unleaded gasoline looks limited.

2.2.3 Co-anti-knock additives

In the late 1950s two laboratories [14, 15] found evidence for the beneficial effect of carboxylic acids (or of materials which will produce such carboxylic acids in the engine) on the octane quality of leaded aromatic gasolines. These materials were called lead extenders or lead appreciators. Many different materials were tested and, at practical concentrations of 3000 ppm or more, benefits in RON of leaded gasoline of two or more numbers were found [15]. The beneficial effects were usually greater for TEL than for TML, although in aromatic fuels TML initially performs better than TEL. A mechanistic explanation [16] of the behaviour of the co-anti-knocks is that they interfere with the agglomeration process of lead oxide particles. Given enough time, the agglomeration process reduces the effectiveness of the lead, and the co-anti-knock can be seen as preserving the active concentration of the lead oxide at the time in the engine cycle when it is needed. This mechanism also explains the effect of too great a concentration of the carboxylic acid, which is found to lose its co-anti-knock effect by tying up all the lead oxide in the form of lead salts.

It does not appear that any commercial use has been made of this class of material. The carboxylic acids would have certain undesirable side effects (e.g. of corrosion) and the more desirable esters are intrinsically much less effective. The final arbiter of cost effectiveness also appears to be unfavourable, and as the lead level decreases, any co-anti-knock benefit naturally will decrease with it.

2.2.4 Ashless anti-knock additives

One way to minimize the effects of undesirable combustion products of anti-knock additives is to make them ashless (e.g. of C, H, O and N in

Gasoline additives influencing combustion processes   51

composition). Indeed, some of the earliest additives of all in the 1920s were aromatic amines, and aniline was in fact used as the standard against which all such materials were compared. In the mid-1950s two studies [17, 18] of amines were made which showed their effectiveness in the fuels then available. In this same period there was an excellent, and still standard, study of the octane quality of scores of hydrocarbon compounds by the API 45 project [19]. Also, in the rather scattered patent literature there have been regular claims of anti-knock activity on behalf of almost all types of compound. Yet after all this activity, lead alkyls remained clearly ahead on cost-effective terms.

As the prospect of legislation against lead became more prominent, another comprehensive experimental study was made by MacKinven [20] and reported in 1974. Over 970 compounds were tested and rated for their anti-knock capability, regardless of cost, toxicity, etc. Four groups of compounds emerged with appreciable activity relative to that of N-methylaniline: aromatic compounds containing nitrogen (as amines, hydrazines and N-nitrosamines), aromatic compounds containing oxygen (as phenols and esters), elemental iodine and aliphatic iodine compounds, and selenium compounds. Of these, none were as much as three times as effective as NMA, which was estimated by MacKinven to be the breakeven point relative to gaining the same octane advantage by refinery processing. Accordingly, the study concluded that none of the materials provided a more economic route to high octane numbers than processing.

Certain oxygenated compounds such as methanol, ethanol, methyl tertiary butyl ether and others have been promoted as lead substitutes, since they have proved to be valuable in helping to achieve the required octane quality for low-leaded and unleaded gasolines. These materials are perhaps more appropriately regarded as blend components than additives, since they are used at relatively high concentrations. They are not considered here since they are discussed in detail in Chapter 5.

Claims still occur in the literature for anti-knock additives, and it is helpful to realize that several possible mechanisms of action of such materials exist. The additives that simulate lead alkyls in offering an immediate improvement in octane quality of the fuel would most likely operate by a similar chemical mechanism. Thus the additive molecule itself or (more likely, and analogously, to lead alkyls) a degradation product generated in the pre-flame reactions of the engine compression stage (or even the cycle before) would act as a free-radical trap in the end-gas zone to retard the chain reaction which leads to premature auto-ignition. However, other mechanisms are possible of a more heterogeneous nature, and these could possibly give rise to beneficial effects by means of deposit-modification over a long period of time, such as after one or more tankfuls of fuel.

2.3 Anti-ORI (octane requirement increase) additives

Octane requirement increase (ORI) is a feature of an ageing engine that has sometimes not been recognised, even though work began in this subject in the 1950s or even earlier, and has been pursued steadily inside several automotive laboratories for a number of years. Yet the fact is that the octane requirement of a modern car increases typically by 3–5 octane numbers over the first 10 000–15 000 miles of its life. After this time the engine stabilizes and no further systematic changes in ORI take place.

The factors which affect ORI are several: work by Benson [21] showed that its magnitude depended on driving pattern, lubricant composition

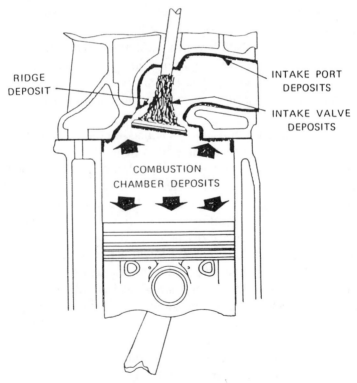

Figure 2.6 Engine deposit locations which cause octane requirement increase.

Gasoline additives influencing combustion processes

and fuel-additive composition but not on lead content of the fuel or the use of ashless lubricant. Other work has shown that lubricant composition is an important factor, and that there may be larger effects with unleaded than with leaded fuel. The origin of the phenomenon lies in the accumulation of engine deposits. Previous work with unleaded fuel showed that all the ORI could be attributed to the growth of combustion chamber deposits and that most of the effect could be attributed to deposits in the end-gas zone. Later work by Graiff [22] showed that in certain vehicles the growth of deposits in the intake-port area (Figure 2.6) could have as great an effect as the combustion chamber deposits. In understanding the role of the deposits various authors have concluded that volumetric effects of deposits in the combustion chamber are much less important than the thermal ones.

It is therefore plausible to look for additives that might affect the ORI of engines by virtue of their effect on the amount or nature of the engine deposits. Two old and two recent examples can be cited of such ORIC (ORI control) additives.

2.3.1 Halogen scavenger additives (see also Section 2.1)

The early use of TEL in gasolines was plagued by the accumulation of combustion chamber deposits; indeed, the very use of lead alkyl additives depended on a solution being found. These deposits gave rise to serious problems of exhaust valve burning and damage to the insulators and the electrodes of the spark plugs. However, they must also have caused a loss in octane requirement performance, and so the scavenger additives that were discovered can also be regarded as the first ORIC additives.

2.3.2 Boron deposit modifier additives

In the mid-1950s Hughes *et al.* [23] pioneered the addition of boron compounds which proved effective with leaded gasolines. It was believed that the boron oxides in the burned gases combine with lead compounds to form borate deposits which are 'unusually friable', and can therefore facilitate mechanical scavenging of the deposits. The result was some protection against surface pre-ignition (see below) as well as some benefit in ORI. The boron additives were believed to be based on the glycol borate family of chemicals.

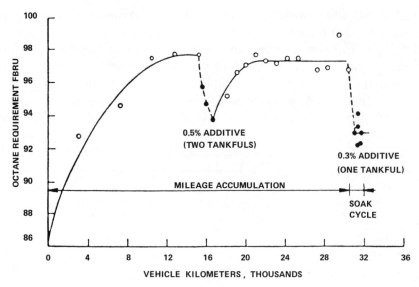

Figure 2.7 Effect of an additive, used intermittently, on the octane requirement of a mileage accumulator vehicle.

2.3.3 Detergent additives used intermittently at a high dose rate

In 1983 Bert et al. [24] described their polyetheramine detergent material which, when used in one tankful of gasoline, was demonstrated to reduce ORI by 30–40 percent of the original engine ORI. The benefit was claimed to last for several thousand kilometres until the deposits re-establish themselves in the engine. The mechanism postulated for the effect was that of a modification to the combustion chamber deposits, although the detergent does also affect deposits elsewhere in the induction system. An example of the results in a vehicle on a mileage accumulation is shown in Figure 2.7. The additive, known as Techron, has been widely used commercially.

2.3.4 A detergent additive used continuously at a low dose rate

In 1985 an oil company in the USA launched a detergent additive package, PDP5400, which claimed not only engine cleanliness benefit but also ORI benefits. The additive was based on the earlier work by Graiff [22], which demonstrated the connection between inlet-port and ridge deposits and ORI. For example, in tests [25] on a 1977 model 4.9-litre engine, out of a total octane requirement increase of 8.6 numbers, 6.9 was attributed to the intake port and valve deposits and 1.7 to the combustion

Gasoline additives influencing combustion processes 55

chamber deposits. At the same time, the intake port deposits appeared beneficial to fuel consumption. The effect of the detergent additive was to control these intake deposits, and thereby confer benefits on ORI without significant loss of fuel economy.

2.4 Anti-run-on additives

Run-on (or afterrun, as it is sometimes known) is an occasional occurrence in modern engines, and can be regarded as a driveability 'nuisance'. It occurs when the engine 'diesels' after the ignition is switched off.

The engine and fuel factors controlling this phenomenon have been well researched and understood [26, 27]. Engine speed at idle was found to be the most important engine factor of all, so throttle-stop settings are of practical importance. As far as the fuel is concerned, octane quality, governed by the RON, is the predominant property. Engines are tuned such that they almost invariably reach a knock limit before they reach a run-on one (although it is conceivable that the reverse could be the case; e.g. in a heavily spark-retarded engine).

The additives that are effective to control this phenomenon are therefore the same that are used for anti-knock purposes (e.g. lead alkyls or others). Similarly, it might be expected that anti-ORI additives would also show anti-run-on capability. The Techron additive was indeed tested [24] in this regard and positive evidence was adduced in a 5.0-litre V8 bench engine with unleaded gasoline. For a given RON level, run-on was found to greatly increase in the presence of equilibrium fuel deposits compared with the clean engine, and was decreased somewhat after use of the octane requirement decreasing additive treatment.

2.5 Anti-pre-ignition additives

In the 1950s as compression ratios first increased up to 9:1 or more the phenomenon of pre-ignition was encountered. The problem showed itself as noisy and sometimes damaging combustion behaviour, and attracted such descriptive titles as 'wild ping', 'rumble' or 'picket-fence rattle'. It was attributed to the presence of growing amounts of combustion chamber deposits arising from the presence of large amounts of lead in the high-octane fuels of the day. These deposits became hot enough to ignite an incoming charge before the moment of spark-ignition, so the event can properly be termed surface pre-ignition. The point of ignition can vary; sometimes it would be the piston top, often the spark plug tip itself. It is a very dangerous phenomenon because of its propensity to lead to runaway damage. However, sometimes the violence of the process is sufficient to detach the offending deposits, in which case the effect is transitory and self-correcting.

The action of the deposits needs further elucidation [16]. Carbonaceous deposits need to be heated to above 700 °C in air before they start to combust and glow. However, the presence of lead salts catalyses this process such that the glow temperature is much reduced, and so these deposits became a more ready source of pre-ignition. It was discovered [28] in the mid-1940s that phosphorus compounds can raise this glow temperature by converting the lead halides and oxyhalides to less catalytic lead phosphate salts. A number of materials were shown to be effective. Two of the most cost-effective additives were tricresyl phosphate (TCP) and cresyldiphenyl phosphate (CDP). The dose rate was typically 0.2–0.5. Theories of phosphorus relative to lead, calculated on the basis that the two materials combine to form lead phosphate. (Thus for a fuel of 1.0 gPb/l the phosphorus content was typically 0.02–0.05 gP/l). These materials were in much commercial use in the mid-1950s, but as lead contents have decreased steadily in the succeeding 30 years, so too has the use of these phosphorus additives.

2.6 Anti-misfire additives (or driveability additives)

Misfire arises when either the spark itself is not adequate or the fuel–air mixture is not able to sustain combustion (or very often in practice a combination of both). Fuel additives have been developed to combat both sources of the problem.

2.6.1 Spark plug anti-foulant additives

In use, deposits can form on the spark plug surface, thus forming an electrical bridge or pathway between the centre electrode and the earthed body. If these deposits have any electrical conductivity then, under certain operating conditions, the spark energy delivered to the combustion gases will decrease enough for misfire to occur. In fact, lead halides and oxyhalides under high operating temperatures do tend to melt and become very conductive, and with high-lead fuels it was a common experience for spark-plug misfire to be encountered under full-throttle acceleration conditions. However, somewhat fortuitously, the same phosphorus additives (TCP or CDP) that were effective in modifying the catalytic lead deposits in the combustion chamber deposits were also

Gasoline additives influencing combustion processes

effective in modifying them in the spark plug [29]. Lead phosphate salts are much less conductive than the lead halides and oxyhalides at high temperatures and, moreover, the use of the phosphorus at the same doping level (0.2–0.5 Theory) as for surface ignition control also provided spark-plug fouling control. The additive was widely marketed in the 1950s and improved driveability and acceleration performance were claimed. The need of such additives has declined as engine combustion chamber and spark plug design have been evolved and as lead contents in gasoline have been reduced.

2.6.2 Spark-aider additives

In any engine there can come a point when the fuel–air mixture is so dilute that the misfire limit is reached. This situation arises, for example, when transient mixture strength variation occurs during cold-start and warm-up, pushing the mixture to the lean limit: it can also happen when exhaust residuals dilute the mixture to a similar point under part-throttle conditions. At these extreme mixture conditions the spark event is just not able to initiate a strong enough initial flame kernel.

In 1986 a new class of additive (called spark-aiders) was commercialized for the first time. This additive has the ability of improving the spark event

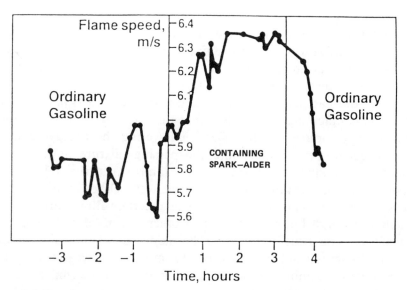

Figure 2.8 The effect of a gasoline containing a spark-aider additive on early flame speed in a Vauxhall Astra engine.

under these borderline conditions. The practical result is measurably improved driveability, especially over cold transient driving conditions.

The additive is based on an organic potassium compound that is soluble in the gasoline, and which at a potassium dosing level of a few ppmw is found to form, and keep on renewing, the deposit layer on the spark plug. This deposit layer has the property of a lower electronic work function, which results in an 'easier' spark that, in turn, delivers more of the spark electrical energy into the gas mixture. The initial flame is therefore stronger, and increases in early flame speed together with improvements in the cycle-by-cycle variability have been demonstrated [30] in an engine equipped with quartz windows and laser diagnostics (Figure 2.8). Practical benefits in vehicle driveability have also been demonstrated in both leaded and unleaded gasolines. Other soluble metallic compounds of Groups I and II metals have also been shown to be effective spark-aiders, to a greater or lesser degree.

2.6.3 Additives for improving fuel distribution

A well-recognized cause of poor driveability in multi-cylinder engines arises from the maldistribution of fuel–air mixture between the cylinders. The poor performance is determined by the onset of misfire in only one of the cylinders, a situation most likely to occur in lean-set carburetted cars driven over cold-start or warm-up conditions.

An additive route to ameliorate this problem was developed [31] in 1972. The additive was a surface-active material called HTA (hydrogenated tallow amine), and consisted of an ashless mixture of non-polymeric amines plus solvent. It worked by coating the surface of the inlet manifold in order to decrease its wettability with gasoline, i.e. the gasoline tends to form high-contact-angle droplets rather than a spread-out layer. Such a low-energy surface has the effect of lowering the energy required for droplets to be entrained into the passing mixture flow, thereby enhancing the mixture uniformity and improving driveability. This improvement was demonstrated in vehicle tests. Other materials were also studied and found to be effective (e.g. certain fluorocarbons and dimethyl silicones), but were rejected for other reasons of cost or adverse side effects. The benefits of the HTA additive on fuel–air mixture distribution should not only give rise to improved driveability performance but also to improvements in fuel economy and HC emissions, and indeed such evidence was obtained.

2.7 Anti-VSR (valve seat recession) additives

A well-known problem of unleaded gasoline is that for certain 'soft' metallurgies of the exhaust valve seat a significant wear problem occurs. The high temperatures of the environment together with the absence of semi-solid or solid-lubricant materials in the combustion products of leaded gasoline make for a very hostile environment. Valve seat recession (or sinkage) can occur in such engines at a great rate; sufficiently large, for example, for the engine to use up all its valve-tappet clearance in a period of, say, 10 000 km if it is driven under severe road conditions.

2.7.1 Lead additives

The presence of lead alkyls as anti-knock additives fortunately leads to good VSR protection: even levels as low as 0.05 gPb/l can be sufficient to control the phenomenon. Indeed, one currently commercially recommended route in a vulnerable vehicle running on unleaded fuel is to use lead as a VSR additive by fuelling every fourth tankful with leaded gasoline (of 0.15 gPb/l). In this way, a fresh layer of the lead salt is put down on the exhaust valve seat. However, such a consumer-dependent route does not seem a very reliable or safe answer, and so research has continued to seek alternative VSR additives.

2.7.2 Phosphorus additives

It has been well known for many years [32] that phosphorus, when added to leaded gasoline, also affords VSR protection. The same materials which are effective as ignition control additive (namely TCP and CDP) are also effective in this respect. The point has also been proved [28] in unleaded gasoline using tritolyphosphate and zinc dialkyldithiophosphate. However, such materials have not been used commercially for this purpose.

2.7.3 Alkaline metal additives

Work in the early 1970s [33] showed a sodium naphthenate additive to be only moderately effective, but more recently a manufacturer has developed [34] a sodium–sulphur-based material (called PowerShield) that is effective in controlling VSR in low lead gasolines. At a dose rate of 1000 ppm of the additive (i.e. 10 ppm sodium, 9 ppm sulphur) the protective effect is claimed to be the equivalent of about 0.13 gPb/l. The manufacturer also claims that the additive is effective in unleaded gasoline, but cannot yet recommend it for use in the USA until an EPA waiver is obtained.

2.7.4 Other materials

In the search for VSR control additives in the early 1970s it was considered that abrasive iron oxides form part of the wear mechanism responsible for VSR. Additive materials were therefore chosen with a number of possible control mechanisms in mind: additives with known extreme-pressure lubricant properties, or with an ability to form ashed products (salts) or an oxide film that could provide a barrier to prevent metal-to-metal contact. Apart from the strongly beneficial effects with lead and phosphorus additives, mildly beneficial effects were noted [33] with sodium, vanadium and zinc, and no effects with such additives as iron, boron or sulphur.

2.8 Additives that improve lubricant performance

Some gasoline additives are designed to function by helping the lubricant after surviving unchanged in the combustion chamber. They do this in two ways. First, when an engine is switched on or off it rotates for a few cycles before or after ignition, and this means that gasoline containing the additive is brought into the combustion chamber and not combusted. This gasoline flows down the bores into the crankcase, diluting the lubricating oil so that any additives can have an effect in the bore/ring areas and in the lubricant itself. Second, if the additives are reasonably stable to oxidation a small percentage of them will survive the combustion process and again reach the bores and the lubricant. Some additives which function by these mechanisms are as follows.

2.8.1 Upper-cylinder lubricants/friction modifiers

Upper-cylinder lubricants were one of the first gasoline additives to be used, and consist of a light mineral oil which was added at concentrations up to 0.5 percent by volume. They were useful in the days when lubricants were relatively simple mineral oils which drained away from the bores when the engine was switched off, so that considerable bore wear could occur during the first mile or so of operation before the oil had properly circulated. They are still used today to some extent, although the need for them to protect against bore wear has vanished in the light of improved lubricant technology. Mineral oils are sometimes a constituent of additive packages, since they can play a part in keeping inlet valves and manifolds free from deposits (see Chapter 1) and, as such, will have some effect in lubricating the bores and piston rings. These oils can be quite viscous, and if present at concentrations above 500 ppm may cause a thickening of the crankcase oil and give rise to excessive combustion chamber deposits.

Today, a more important reason for wishing to keep the bores well lubricated is to minimize friction, since a major part of engine friction occurs in the piston ring area. Additives which are particularly effective in this boundary layer regime can be used and can give fuel economy benefits of up to 3 percent. Corresponding benefits in power output can also be claimed. A number of more sophisticated additives than mineral oils have been considered for this purpose, and these include molybdenum-based additives and also surfactant additives known as friction modifiers. These have polar groups which attach themselves to the bores and oleophobic groups which maintain an oily film that does not drain away.

2.8.2 Anti-wear additives

These additives are restricted in their anti-wear properties to preventing corrosive wear in the bore area, and at least one oil company mounted an advertising campaign some years ago on the basis of using such an additive in their gasoline. The theory is that the corrosive gases formed during combustion condense on the bores when the engine is switched off, and can cause significant corrosion if the vehicle is left unused for several hours. When the engine is restarted this rust is swept away by the piston rings, thus presenting a clean surface for more rust to appear at the next shutdown, and causing at the same time a considerable amount of bore wear. The additive used was a surfactant anti-corrosion additive which attached itself to the bores and gave a monomolecular protective film against rust.

Present-day lubricants are so effective that any benefits for this type of additive are now quite small.

2.8.3 Anti-sludge additives

This is another example of an additive type that, because of advances in lubricant design, no longer has a large enough effect to justify using it and advertising its benefits, although in situations where simple mineral oils are used it could be worthwhile. The additive consists of a surfactant dispersant which, as it finds its way into the lubricant, continually 'tops up' the dispersant level in the lubricant and so reduces the possibility of crankcase sludge forming. At one time the dispersants used in lube oils were less stable than current ones and were gradually destroyed by oxidation. Dispersants are very commonly used in gasolines today (see Chapter 1), but this potential benefit is rarely advertised.

2.9 Miscellaneous additives

The patent literature if full of inventors' additives with a wide variety of claims, and a review of this disparate source cannot be given here. The fact that few (if any) of these have come to market probably speaks for itself.

There have been similar claims for additives promoted by smaller commercial companies, many of these exaggerated and often of the 'cure-all' variety. Such additives seldom stay in the market for long, although it should be said that it can be a very difficult and expensive matter to prove or disprove claims made in this way.

Lastly, there have been certain additives developed by the more important oil or additive companies that have been produced for a short time (e.g. combustion catalysts) or for which research findings have been fairly well documented but which, for some reason, have not been commercialized.

2.10 Future prospects

Research will doubtless continue for additives which can produce measurable benefits in modern engines and gasolines, and opportunities will be provided by future development in these areas. Constraints and opportunities will be brought about by the growth of environmental concern: ensuring that combustion products of the additive are harmless is an example of the first; providing fuel saving over the lifetime of the engine is an example of the second.

An early researcher into anti-knock, T. A. Boyd of General Motors, when reviewing [9] the development of lead alkyls in the 1950 Horning Award SAE Lecture said rather presciently:

> There is at least another 30% gain in economy to be had, whenever the fuels for it can be provided in a commercially practicable manner. And even this is probably not the end at all.

2.11 References

1. H. R. Ricardo and J.G. Hempson, *The High-Speed Internal Combustion Engine*, 5th ed., Blackie, Glasgow (1968).
2. C. F. Kettering, 'More efficient use of fuel', *S.A.E. Trans.*, **4**, 263 (1919).
3. T. Midgely and T. A. Boyd, *JSAE*, **15**, 659 (1920).
4. T. Midgley and T. A. Boyd, 'Chemical control of gaseous detonation with particular reference to internal-combustion engines', *Ind. Eng. Chem.*, **14**, 894 (1922).
5. I. C. H. Robinson, Chapter 3 in *Technology of Gasoline*, ed. by E. G. Hancock, *Critical Reports on Applied Chemistry*, Vol. 10 (1985).
6. T. J. Russell, *Motor Gasoline-antiknock Additives*, Associated Octel Company Report ELP 87/10 (1987).
7. Octel Compound Data, Associated Octel Company Booklet 29/72 (1981).
8. F. J. French, *Criteria for Selection of Lead Anti-knock Compound*, Associated Octel Company Report, SIP. 23 (1976).
9. T. A. Boyd, 'Pathfinding in fuels and engines', *S.A.E.Q. Trans.*, **4**, 182 (1950).
10. J. E. Brown and W. G. Lovell, 'A new manganese antiknock', *Ind. Eng. Chem.*, **50**, 1547 (1958).
11. J. E. Faggan, J. D. Bailie, E. A. Desmond and D. L. Lenane, 'An evaluation of manganese as an anti-knock in unleaded gasoline', SAE Paper No. 750925 (1975).
12. J. B. Moran, 'The environmental implications of manganese as an alternative anti-knock', Society of Automotive Engineers, Warrendale, SAE Paper No. 750926 (1975).
13. J. D. Benson, R. J. Campion and L. J. Painter, 'Results of a Co-ordinating Research Council MMT field test program', Society of Automotive Engineers, Warrendale, SAE paper No. 790706 (1979).
14. S. R. Newman, K. L. Dille, R. Y. Heisler and M. F. Fontaine, 'Tertiary-butyl acetate—an octane improver for leaded gasoline', Society of Automotive Engineers, Warrendale, SAE Paper No. 127U (1959).
15. W. L. Richardson, M. R. Barusch, W. T. Stewart, G. J. Kautsky and K. R. Stone, 'Carboxylic acids extend the anti-knock effectiveness of tetraethyl lead', *J. Chem. Eng. Data*, **6**, 309 (1961).
16. M. R. Barusch and J. H. Macpherson, 'Engine fuel additives', Chapter 10 in *Advances in Petroleum*, Vol. 10, Interscience Publishers/John Wiley and Sons, Chichester (1965), p. 457.
17. J. E. Brown, F. X. Markeley and H. Shapiro, *Ind. Eng. Chem.*, **47**, 2141 (1955).
18. J. A. Brennan, J. J. Giammaria and E. A. Oberright, Am. Chem. Soc. Meeting General Report. No. 1, Vol. 2, p. 211 (April 1957).
19. API Research Project 45, *Knocking Characteristics of Pure Hydrocarbons*, ASTM Special Tech. Publication No. 255 (1958).
20. R. Mackinven, *A Search for an Ashless Replacement for Lead in Gasoline*, Jaghrestagung, DGMK, West Germany (1974).
21. J. D. Benson, 'Some factors which affect octane requirement increase', Society of Automotive Engineers, Warrendale, SAE Paper No. 750933 (1975).

22. L. B. Graiff, 'Some new aspects of deposit effects on engine octane requirement increase', Society of Automotive Engineers, Warrendale, SAE Paper No. 790938 (1979).
23. E. C. Hughes, P. S. Fay, L. S. Szako and R. C. Tupa, *Ind. Eng. Chem.*, **98**, 1858 (1956).
24. J. A. Bert, J. A. Gething, T. J. Hansel, H. K. Newhall, R. J. Peyla and D. A. Voss, 'A gasoline additive concentrate removes combustion chamber deposits and reduces vehicle octane requirement', Society of Automotive Engineers, Warrendale, SAE paper No. 831709 (1983).
25. L. B. Graiff, 'Some new aspects of deposit effects on engine octane requirement increase and fuel economy, Society of, Automotive Engineers, Warrendale, SAE paper No. 790938 (1979).
26. W. S. Affleck, P. E. Bright and R. J. Ellison, 'Run-on in gasoline engines', *I. Mech. E. Proceedings*, **183**, Part 2A (1968–9).
27. J. D. Benson, 'Factors which affect afterrun in gasoline engines', Society of Automotive Engineers, Warrendale, SAE Paper No. 720085 (1972).
28. J. M. Campbell US Patent No. 2,405,560 (1946).
29. R. J. Greenshields, *SAE Trans.*, **61**, 3 (1953).
30. D. R. Blackmore, L. B. Graiff, G. A. Harrow, J. M. Jones, G. T. Kalghatgi and R. Miles, 'Development of a novel gasoline additive package—laboratory test work', Paper C311/86, I. Mech. E. Conference on Petroleum Based Fuels and Automotive Applications (1986).
31. A. A. Zimmerman, L. E. Furlong and H. F. Shannon, 'Improved fuel distribution—a new role for gasoline additives', Society of Automotive Engineers, Warrendale, SAE paper No. 720082 (1972).
32. W. Giles, 'Valve problems with lead-free gasoline', Society of Automotive Engineers, Warrendale, SAE Paper No. 710368 (1971).
33. R. F. Barker, 'Some aspects of the problem of valve-seat wear with non-leaded gasolines', Inst. Mech. Eng. Paper. No. C291/73 (1973). Also at the First European Tribology Conference, London, 1973.
34. R. C. Tupa, 'Today's gasoline concerns—injector plugging and valve seat wear', NPRA Paper AM-86-21 (1986).

3 Additives influencing diesel fuel combustion

T. J. Russell
The Associated Octel Company Ltd, Engine
Laboratory, Watling Street, Milton Keynes MK1 1EZ

3.1	The diesel combustion process	65
3.1.1	Compression ignition	65
3.1.2	Engine design	67
3.1.3	Fuel requirements	68
3.1.4	Exhaust emissions	70
3.2	**Cetane number improvers**	**71**
3.2.1	Cetane number	71
3.2.2	The cetane quality of diesel fuels and diesel fuel components	74
3.2.3	Octyl nitrate cetane improvers	76
	(a) Properties and application	76
	(b) Effectiveness and economics	81
	(c) Effect on engine performance	81
3.2.4	The future potential of other cetane improvers	85
3.3	**Diesel detergents**	**88**
3.3.1	Diesel engine nozzle coking	88
3.3.2	Additives used as diesel detergents	91
3.3.3	Performance benefits from diesel detergents	92
	(a) The effects of reduced nozzle fouling	92
	(b) Other benefits of detergent use	95
3.4	**Diesel smoke suppressants**	**97**
3.4.1	Smoke formation in diesel combustion	97
3.4.2	Additives which reduce diesel smoke	99
3.4.3	Future control of diesel engine exhaust particulates	102
3.5	**Acknowledgement**	**103**
3.6	**References**	**103**

3.1 The diesel combustion process

3.1.1 Compression ignition

Combustion in the diesel engine is achieved by compression ignition. Rapid compression of air within the cylinders generates the heat required

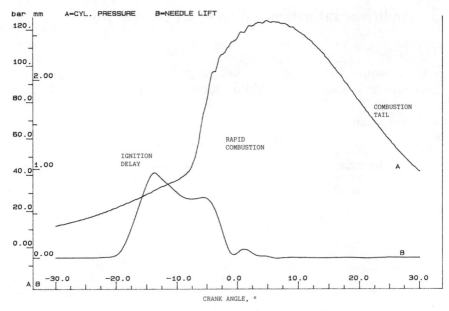

Figure 3.1 Diesel engine combustion cycle.

to ignite the fuel as it is injected. The combustion cycle, as illustrated in Figure 3.1, can be considered in several phases [1]. These include ignition delay, an initial phase which follows compression, when in-cylinder pressures have been considerable increased, temperatures in excess of 500 °C achieved and fuel injected. This phase comprises both a physical delay attributed to fuel atomization, vaporization and mixing, plus a chemical delay during which the reactions of combustion proceed only slowly. This is followed by rapid combustion, a phase in which the fuel–air mix prepared during the delay phase ignites in uncontrolled combustion. The rate and effect of combustion in this second phase is therefore very dependent upon the length of ignition delay and quantity of fuel introduced during the delay pariod.

Subsequent combustion is then, to some extent, regulated by the rate of fuel injection as well as by mixing and diffusion processes. The major controlling factor is, however, the need for the fuel to find oxygen. Ultimately, a combustion tail can be considered to exist, and this is the phase after all fuel has been injected when combustion continues at a reducing rate until all fuel and oxygen have been used.

3.1.2 Engine design

The diesel engine must achieve compression temperatures and pressures which are adequate to enable starting from cold and to give acceptable smooth combustion when running. In high-speed automotive diesel engines this is obtained by the use of compression ratios of between 12:1 and 23:1, depending on cylinder size, combustion system and whether the engine is turbocharged. It is the typically high compression ratio of the diesel engine which dictates the need for its general mechanical robustness and also leads to its high indicated efficiency characteristics.

After compression ratio, the main factors affecting combustion in the diesel engine are combustion chamber design and fuel injection characteristics. The time from start of fuel injection to end of combustion is very small, and during this a fuel–air mixture has to be formed which will ignite within a reasonable delay period. Good mixing of fuel and air is essential and adequate oxygen has to be available for combustion to be completed early in the expansion stroke.

In general, two combustion chamber forms are principally used in automotive diesel engine applications. These are the direct injection system as illustrated in Figure 3.2 and the pre-chamber, indirect injection system shown in Figure 3.3 [2].

With the deep bowl direct injection system the required fuel–air mixing is achieved by direct injection of fuel into an open cylinder into which air

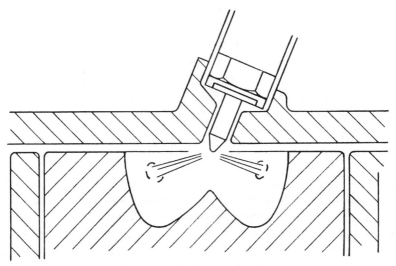

Figure 3.2 Direct injection combustion system.

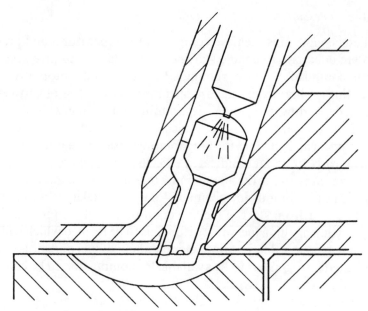

Figure 3.3 Indirect injection combustion system.

has been introduced with a high degree of rotational movement. The air movement is initiated during the induction stroke by correctly sited and shaped inlet ports which causes the air to swirl in the cylinder. The swirl continues as the piston rises during compression.

In contrast, in the pre-chamber indirect injection system air is forced into a pre-chamber during the compression stroke. Fuel is injected into the turbulent air, and because the chamber is not particularly sensitive to atomization pintle-type nozzles are often used. After ignition the pressure in the chamber forces the burning mixture through a narrow passage or passages into the cylinder, where it mixes with air to complete combustion.

3.1.3 Fuel requirements

A principal requirement of fuels for high-speed diesel engines is good ignition quality. Fuels of poor ignition quality can lead to extended ignition delay and result in the phenomenon of 'diesel knock'. Excess fuel is injected into the combustion chamber during the delay period and in combustion causes more rapid and greater rises in cylinder pressure. This results in increased engine stress and noise together with possible losses in power and fuel economy. Changes in the ignition quality of fuels can result in rougher engine running, particularly at light loads, and can also have effects on engine exhaust emissions. In addition, in extreme circumstances the ability of a diesel engine to fire and run will be linked to the ignition quality of available fuels.

The most universally accepted measure of the ignition quality of diesel fuels is cetane number. As shown in Figure 3.4, the self-ignition character-

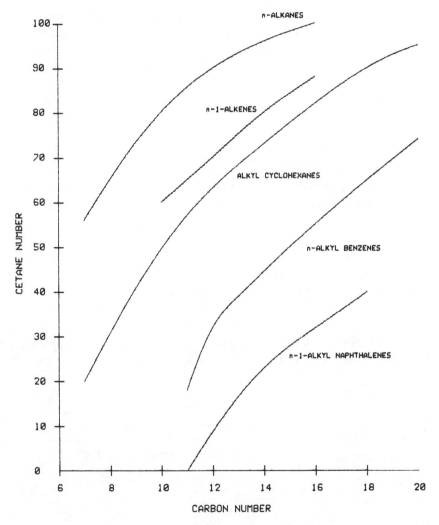

Figure 3.4 Cetane number of pure hydrocarbons.

istics of hydrocarbons vary markedly with both boiling point and chemical structure [3].

Traditionally, a distillate fraction from crude oil within the 180–370 °C boiling range has been used as diesel fuel. For the majority of crude oils this fraction contains a high proportion of paraffinic hydrocarbons and is of excellent ignition quality. The increased use of cracked components in diesel fuel blending is, however, leading to a general lowering of diesel fuel ignition quality. The aromatic hydrocarbons present in these components have a particularly poor ignition quality.

The physical characteristics of a diesel fuel are vitally important with respect to the production of a fuel suitable for commercial use. Ignition quality is of prime importance when considering combustion, but other properties such as volatility, viscosity, hydrocarbon composition and stability cannot be overlooked if optimum engine performance is to be achieved.

3.1.4 Exhaust emissions

In general, with the exception of particulate matter, emissions from the diesel engine have been considered low when compared with the spark-ignition engine. Comparative data are given in Table 3.1 and show very low emissions of carbon monoxide and hydrocarbons.

Increasing attention is, however, being focused on vehicle exhaust emissions, specifically on particulate and oxide of nitrogen levels. Proposed legislation is most severe in the USA and the use of particulate traps on heavy and light duty diesel engines is envisaged.

Within Europe no limit on particulate emissions from heavy-duty diesels is as yet proposed and the only control remains the steady-state full-load smoke test. Controls on particulates are, however, proposed for light-duty vehicles such as cars and small trucks [4].

Table 3.1 Exhaust emissions—European passenger cars registered 1984

	Exhaust emissions g/test ECE 15			
Model	Hydrocarbons	Carbon monoxide	Oxides of nitrogen	Particulates
Gasoline (1.3-litre)	13.94	110.27	5.54	—
Diesel (1.6-litre IDI)	1.87	6.88	2.14	0.73

3.2 Cetane number improvers

3.2.1 Cetane number

As previously stated, the most universally accepted measure of the ignition quality of diesel fuels is cetane number. The standard method for determining the cetane number of a diesel fuel is the ASTM D613 CFR engine technique [5]. In this procedure the cetane number of a diesel fuel is determined by comparing its ignition quality with two reference fuel blends of known cetane numbers under standard operating conditions. This is done by varying the compression ratio for the sample and each reference fuel to obtain a fixed delay period between the start of injection and ignition. The compression ratio for the sample is bracketed by reference fuel blends, which differ by less than five cetane numbers, and the rating of the sample is calculated by interpolation.

The method uses a single-cylinder CFR engine, operated under the basic conditions given in Table 3.2. The cetane number scale is based on two primary reference fuels. These are n-cetane (n-hexadecane), which has a cetane number of 100, and heptamethyl nonane, which has a cetane number of 15. In practice, however, in view of the high cost of these materials, secondary reference fuels blended by Phillips Petroleum are used.

Table 3.2 CFR engine operating conditions

Operating requirement	Method ASTM D613
Compression ratio	Variable 7:1 to 28:1
Ignition delay (crank angle degrees)	Fixed 13°
Start of injection (crank angle degrees)	13° BTDC
Start of combustion	TDC
Intake air volume	Constant
Intake air temperature (°C)	65 ± 1
Engine speed (rpm)	900 ± 9
Fuel flow (ml/min)	13 ± 0.2
Engine load (Nm)	Not specified
Water jacket temperature (°C)	100 ± 2
Oil temperature (°C)	57 ± 8
Recommended reference fuels	
Primary	n-cetane/heptamethyl nonane
Secondary	Phillips Petroleum T and U fuels
Calculation technique	Bracketing pair
Recommended reporting interval	Nearest integer

Table 3.3 Precision of cetane number determination

Cetane number	Repeatability	Reproducibility
40	0.6	2.5
44	0.7	2.6
48	0.7	2.9
52	0.8	3.1
56	0.9	3.3

As a quality control procedure, cetane number measurement leaves much to be desired. The ASTM test method is time consuming, requires skilled operators and, as shown in Table 3.3, is of poor precision.

In view of these shortcomings, prediction equations known as cetane indices have found widespread use in the monitoring and control of diesel fuel ignition quality. The index system currently used (IP 364) provides a rapid means of assessing the ignition quality of fuels using typically available density and distillation data [6]. Cetane index can be calculated using the formula given in Table 3.4 or more conveniently estimated from the nomograph shown in Figure 3.5.

Although index systems can prove extremely useful, care must be taken with their use if misleading results are to be avoided. The cetane index is known to be unsatisfactory with fuels, with unexpected interrelationships between density and distillation mid-boiling point. In addition, as discussed in more depth later, the calculated cetane index does not reflect the effect of cetane improvers on diesel fuel cetane quality.

Various attempts to improve the accuracy of cetane index prediction systems have been made and more complex equations of the type given in Table 3.5 are to be proposed for use by the ASTM and the Institute of Petroleum. This equation is again based on density and distillation properties of the fuel and could well have many of the shortcomings of previous index systems. The more fundamental approach of Glavincevski

Table 3.4 Calculated cetane index of diesel fuels (IP 364/83)

$$\text{Cetane index} = 454.74 - 1641.416D + 774.74D^2 - 0.554B + 97.803 (\log B)^2$$

where D = density at 15 °C
B = mid-boiling temperature (°C) corrected to standard barometric pressure

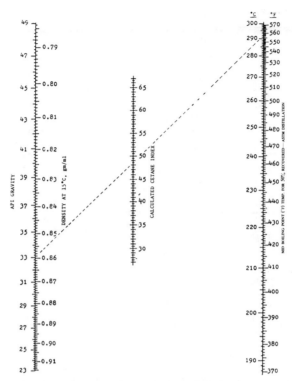

Figure 3.5 Nomograph for calculated cetane index. (*Source*: Institute of Petroleum [6].)
Note: The Calculated Cetane Index equation represents a useful tool for *estimating* cetane number. Due to inherent limitations in its application, Index values may not be a valid substitute for ASTM cetane numbers as determined in a test engine.

Table 3.5 Calculated cetane index of diesel fuels: ASTM four-variable equation

$$\text{Cetane index} = 45.2 + 0.0892(T_{10}N) + 0.131(T_{50}N) + 0.0523(T_{90}N)$$
$$+ 0.901B(T_{50}N) - 0.420B(T_{90}N) + 4.9 \times 10^{-4}(T_{10}N)^2$$
$$- 4.9 \times 10^{-4}(T_{90}N)^2 + 107B + 60B^2$$

where $T_{10}N = T_{10} - 215$
$T_{50}N = T_{50} - 260$
$T_{90}N = T_{90} - 310$

when T_{10}, T_{50}, T_{90} are temperatures at 10%, 50% and 90% volume distilled

and $B = (\exp(-3.5\,DN)) - 1$

when DN = density at 15 °C (kg/l) — 0.85

Table 3.6 Diesel fuel cetane number specifications and typical quality

Country	Cetane number/index specification	typical ranges (min.)	(max.)	Average
Australia	45/-	51.3	54.1	52.4
Denmark	-/50	45.8	48.3	47.3
Finland	-/45	47.8	49.7	49.2
France	-/48	47.0	54.3	50.6
UK	50/-	50.5	54.2	52.8
Italy	47/48	48.8	53.1	50.5
Japan	45/-	51.8	57.1	54.7
Spain	50/45	51.4	53.6	52.8
USA	40/-	42.0	55.5	45.3
West Germany	45/-	43.6	56.6	51.2

Sources: Octel European Diesel Fuel Survey, 1985 and 1986 [8]
Paramins Worldwide Diesel Fuel Quality Survey, 1985 [9]
NIPER, Diesel Fuel Oils, 1987 [10]

[7], which introduces more specific composition data on the fuel, appears applicable to a wider range of products but it is of questionable relevance to the refiner.

3.2.2 The cetane quality of diesel fuels and diesel fuel components

Cetane number or cetane index requirements are present in most diesel fuel specifications but, as illustrated in Table 3.6, significant variation in the quoted values is evident [8, 9, 10]. Within Europe the cetane quality of diesel fuel has traditionally been high, and importance is attached to the performance benefits this offers in terms of:

1. Improved cold starting
2. Reduced smoke emission during warm-up
3. Reduced noise
4. Reduced fuel consumption and exhaust emissions
5. Improved engine durability

The scope of refiners to produce diesel fuel of high cetane quality varies significantly with the types of crude oil processed and the process units available within the refinery. In general, throughout the world the conversion refinery utilizing catalytic cracking predominates, and the diesel fuel blending components typically available are shown in Tables 3.7 and 3.8, the effect of crude source on light gas oil quality being

Additives influencing diesel fuel combustion 75

Table 3.7 Effect of crude source on diesel fuel blending component quality

Property \ Crude source	Light gas oil ex Kuwait crude	Light gas oil ex Forties crude	Light gas oil ex Nigerian crude
Density (kg/l at 15 °C)	0.8517	0.8558	0.8785
Viscosity (cSt at 40 °C)	4.5	4.6	4.5
Cloud point (°C)	4	6	−8
Cold filter plugging point (°C)	−4	1	−11
ASTM distillation			
50% distilled at °C	304	294	283
Sulphur (% wt)	1.40	0.19	0.13
Cetane number (ASTM D613)	54.1	52.2	40.9

illustrated by the data in Table 3.7 and the effect of component type by the data in Table 3.8.

Trends in the production and use of petroleum products are dictating that increased quantities of the lower cetane quality components are used in diesel fuel blending. Projections are that, without the use of additives, marked falls in the cetane quality of diesel fuels could occur. In this situation cetane improvers are invaluable, as they provide refiners and blenders with a simple and effective means of achieving operational flexibility on a day-to-day basis.

Current experience would suggest that cetane number improvers are being increasingly used to:

1. Upgrade diesel fuel quality to meet specification requirements in

Table 3.8 Effect of component type on diesel fuel blending component quality

Property \ Component type	Kerosene ex North Sea crude	Light gas oil ex North Sea crude	Light cracked gas oil ex North Sea crude	
			Non-hydro-treated	Hydro-treated
Density (kg/l at 15 °C)	0.8011	0.8558	0.9613	0.9294
Viscosity (cSt at 40 °C)	1.2	4.6	3.1	3.0
Cloud point (°C)	−58	6	−11	−11
Cold filter plugging point (°C)	−58	1	−9	−9
ASTM distillation				
50% distilled at °C	190	294	276	273
Sulphur (% wt)	0.03	0.19	1.35	0.24
Cetane number (ASTM D613)	41.3	52.2	21.0	24.1

conversion refineries where there is a requirement to use increasing quantities of cracked components in diesel fuel production
2. Provide the flexibility required to process significant quantities of low cetane naphthenic crude oils when geographic or economic conditions dictate this
3. Upgrade diesel fuel quality to give the premium grade products now being marketed by many oil companies in certain markets throughout the world

3.2.3 Octyl nitrate cetane improvers

(a) Properties and application Cetane improvers are compounds which readily decompose to give free radicals and thus enhance the rate of chain initiation in diesel combustion. They promote fast oxidation of fuels and thus improve their ignition characteristics. Chemical compounds such as alkyl nitrates, ether nitrates, dinitrates of polyethylene glycols and certain peroxides are known cetane improvers. In general, however, in view of their low cost and ease of handling, most commercial significance has been attached to different primary alkyl nitrates.

Various products have been marketed at different times, and these have essentially been based on the following compounds or their isomers:

1. Iso-propyl nitrate
2. Iso-amyl nitrate
3. Iso-hexyl nitrate
4. Cyclohexyl nitrate
5. Iso-octyl nitrate

The weak $RO-NO_2$ bond in each of these molecules provides the available source of the free radicals required to enhance diesel combustion.

The relative effectiveness of certain of these alkyl nitrates and other cetane improvers is illustrated in Figure 3.6. Relationships proposed by Li and Simmons [11] suggest that the relative performance of these additives will remain identical in all fuels and at all additive concentrations. The iso-octyl nitrate product, in view of its good response and low production costs, is the most cost-effective additive and is now almost exclusively used in all commercial applications with hydrocarbon-based fuels.

Octyl nitrate cetane improvers are marketed by major additive companies, including Associated Octel, DuPont, Ethyl Corporation,

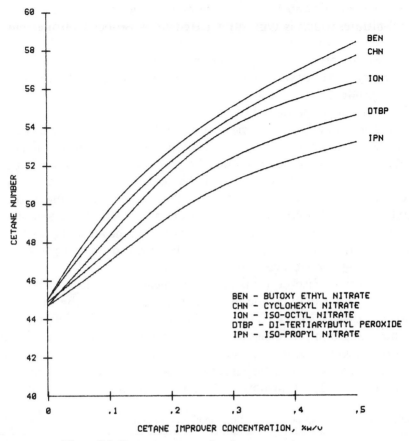

Figure 3.6 Cetane response of various cetane improvers.

Table 3.9 Iso-octyl nitrate cetane number improver: typical composition and properties

Purity	
Total octyl nitrates	99%
Water content (% w/w)	0.05 maximum
Acidity as HNO_3 (% w/w)	0.003
Appearance	
Liquid, from colourless to yellow	
Physical properties	
Density at 20 °C	0.963
Flash point (Pensky closed cup)	80 °C
Pour point	−30 °C
Viscosity centistokes 0 °C	3.00
20 °C	1.70

Exxon Chemicals and Lubrizol. The major product available is 2-ethyl hexyl nitrate, which is typically referred to as iso-octyl nitrate. This is produced by nitration of the commodity alcohol, 2-ethyl hexanol. The mixed isomer products of equivalent performance but containing a range of primary octyl nitrates have now largely been replaced by this single-isomer product.

The typical properties of iso-octyl nitrate cetane improver are summarized in Table 3.9. It is a clear liquid of low viscosity which is easily miscible in diesel fuels. The additive poses no significant handling problems and in general should be treated as a high flash point hydrocarbon liquid. It is non-flammable and has low toxicity.

Cetane improvers are characteristically thermally unstable, and iso-octyl nitrate will decompose in an exothermic reaction if heated above 115 °C. Although this causes no problems during the normal distribution and use of the product, attention must be given to this property in the design of suitable storage facilities.

The addition of iso-octyl nitrate has no measurable effect on diesel fuel properties other than cetane number. This is illustrated by the data in Table 3.10, which show that even at additive concentrations of 0.5 percent v/v no changes in properties such as distillation characteristics, density, cold flow properties, flash point or viscosity were evident. In common with other alkyl nitrate cetane improvers, iso-octyl nitrate does, however, interfere with the Ramsbottom and Conradson carbon residue test

Table 3.10 Effect of iso-octyl nitrate on diesel fuel quality

Iso-octyl nitrate concentration (% v/v)	Zero	0.10	0.20	0.50
Density at 15 °C (kg/l)	0.8516	0.8516	0.8516	0.8516
Viscosity at 40 °C (cSt)	2.7	2.7	2.7	2.7
Cloud point (° C)	−9	−9	−9	−9
Cold filter plugging point (° C)	−8	−8	−8	−8
Pour point (°C)	−27	−27	−27	−27
Flash point (°C)	66	66	65	66
ASTM distillation:				
IBP (°C)	167	167	166	166
10% recovered at °C	191	192	191	191
30% recovered at °C	226	226	226	226
50% recovered at °C	260	260	260	260
70% recovered at °C	297	297	297	297
90% recovered at °C	345	345	346	346
FBP (°C)	374	375	374	374
Cetane number (ASTM D613)	44.6	49.0	51.3	55.0

procedures. Both methods show increased residue in the presence of cetane improver. No adverse effects on vehicle performance exist and each test procedure specifies that carbon residue determination should be undertaken on the base fuel prior to addition of cetane improver, i.e.

NOTE 3—Also, in diesel fuels the presence of amyl nitrate or alkyl nitrate causes erroneously high values. Accordingly, this method is applicable only to base fuels without this additive.

The presence of iso-octyl nitrate in diesel fuel can be determined using the ASTM D4046 test procedure. This analysis is based on hydrolysis of the nitrate followed by nitration of m-xylenol with the nitric acid liberated. The nitroxylenol formed is extracted from the reaction mixture and reacted with sodium hydroxide to form a yellow salt. The colour is

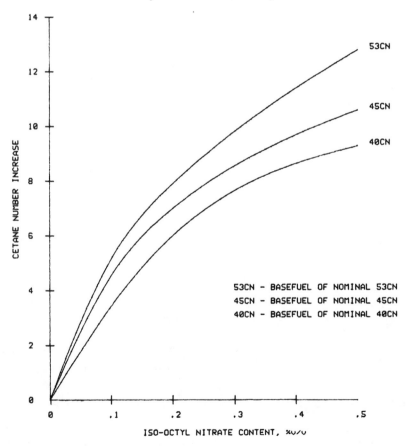

Figure 3.7 Iso-octyl nitrate response in diesel fuels of differing base cetane numbers.

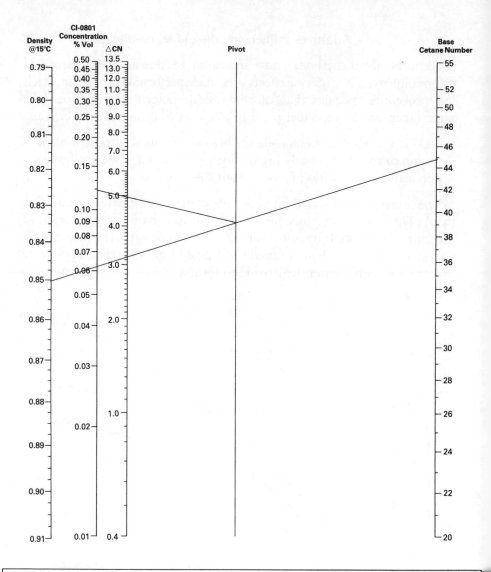

METHOD OF USE
1. Identify the base fuel Cetane Number and density on the appropriate axes and draw a line joining these two points.
2. Where this line intersects the pivot line, either
    a) connect the pivot point to the Cetane Number increase ($\triangle$CN) that is desired and extend the line to the CI-0801 axis to predict the additive concentration, or
    b) connect the pivot point to the chosen CI-0801 concentration and read back along the line to the predicted Cetane Number increase.

EXAMPLE

Base Cetane Number 45, density @ 15°C 0.85, desired Cetane Number increase 5.0. Therefore, the CI-0801 concentration required is 0.12% volume.

Figure 3.8 Nomograph for the calculation of cetane improvement from CI-0801 additions. (*Source*: The Associated Octel Company Ltd.).

measured spectrophotometrically and the concentration of octyl nitrate determined by reference to a prepared calibration curve.

Alternatively, procedures based on infra-red spectroscopy can be used as a more rapid indicator of the presence of iso-octyl nitrate, the absorbance band at 1640 cm^{-1} being a clear indication of the presence of alkyl nitrate compounds in the diesel fuel.

(b) Effectiveness and economics The performance of cetane improver additives is very dependent upon base fuel quality. Typically, four to six cetane numbers would be expected from 0.1 percent v/v iso-octyl nitrate, but, as illustrated in Figure 3.7, this value is closely linked to the original cetane quality of the fuel.

In addition to cetane number, a range of other factors such as hydrocarbon composition and distillation boiling range are known to affect cetane improver performance. Many equations correlating fuel properties with cetane improvement from iso-octyl nitrate exist. The nomograph given in Figure 3.8 presents one of these equations in an easy-to-use form.

It is important to note that cetane improvers are less effective in the low-cetane fuels which are most in need of cetane improvement. Nonetheless, meaningful cetane gains can be obtained and significant economic and performance benefits can be attributed to these.

Although the economic benefits of cetane improver use vary for different refinery configurations and product specifications, requirements and price structure, some general comment is necessary. If, therefore, as exists in many refineries today, cetane improvers are considered as a direct means of upgrading low-cetane number cracked components, to higher value products, then a relationship such as that given in Figure 3.9 can be proposed. Cetane improvers show obvious cost benefits at price differentials typically expected between gas oils and fuel oils.

(c) Effect on engine performance The remaining area of interest concerning the use of cetane improvers is their effect on vehicle performance. Much debate on the importance of cetane number as a fuel specification and also the possible difference between natural (hydrocarbon-derived) and chemical (additive-derived) cetane numbers exists. In discussing this particular topic the varying requirements of different engines cannot be overlooked, and certain engine designs will certainly be more sensitive to fuel cetane quality than others.

Cetane number cannot adequately predict everything concerning the

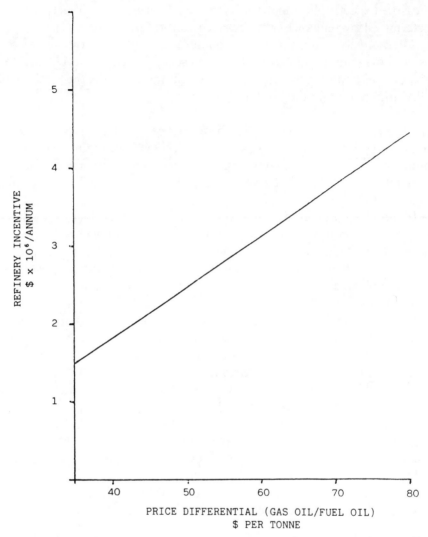

Figure 3.9 Economic incentive to upgrade cat. cracker gas oil to automotive gas oil at constant cetane number, 100 000 B/D conversion refinery.

combustion performance of a fuel and can be misleading with certain potential non-hydrocarbon alternatives. It is, however, a most important indicator of ignition quality for hydrocarbon-derived fuels. High cetane number shortens ignition delay to provide smoother rates of cylinder pressure rise with lower peak pressure and less engine noise, improved cold starting and reduces white smoke emissions during vehicle warm-up.

Additives influencing diesel fuel combustion 83

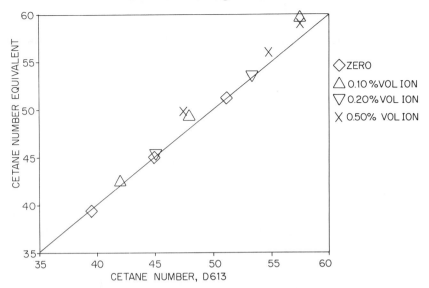

Figure 3.10 Relationship between cetane numbers from ignition delay and ASTM D613 for fuels containing cetane number improver at different concentrations.

Reductions in cetane number can result in increased exhaust emissions.

Recent cooperative work into diesel combustion has confirmed the link between fuel cetane number and engine ignition delay [12]. This showed that fuel cetane number did not universally rank ignition delay in all engines but concluded that the underlying trend in most engines was for ignition delay to decrease as fuel cetane number increased. This trend was also evident for fuels containing iso-octyl nitrate cetane improver and confirms the data given in Figure 3.10. A direct correlation between cetane number equivalent as determined by ignition delay measurements on a modified multicylinder engine and fuel cetane number existed with fuels containing up to 0.5 percent v/v iso-octyl nitrate. No discrimination between natural or chemical cetane quality was therefore evident.

An area of particular concern with cetane improvers is their effect on the cold-start performance of diesel fuels. It has been suggested that additives which boost cetane number as measured by the ASTM CFR engine procedure (or any other engine at normal running temperatures) may not necessarily give improved cold-start or smoke-clearing characteristics. Recent reported work on this topic [13] does not fully support this opinion, and shows that significant benefits can be attributed to the use of cetane improvers under cold-start conditions.

This can be illustrated by reference to the work of Sutton [14]. The

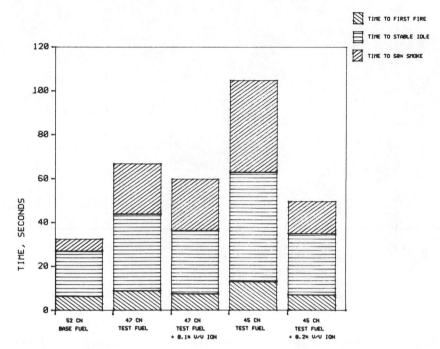

Figure 3.11 The effect of cetane number on cold starting at −10 °C (4.0-litre direct injection engine).

starting characteristics of an indirect injection and a direct injection engine were evaluated at low temperature using fuels of different cetane number. The results are summarized in Figures 3.11 and 3.12, and illustrate the performance of the fuels in terms of:

1. Time to first fire
2. Time to stable idle (800 rpm)
3. Time to reduce white smoke to 50 percent obscurity

The work was completed at −10 °C in a cold chamber using stringently controlled conditions and the data presented are the average of various tests.

In terms of additive performance it is evident that in the direct injection engine most of the increases which result from reduced cetane quality are restored by correct use of cetane number improver. This is particularly noticeable in comparisons of the performance of the 45 CN fuel with and without octyl nitrate. Similar conclusions can be made concerning smoke clearance with respect to the test data for the indirect injection engine.

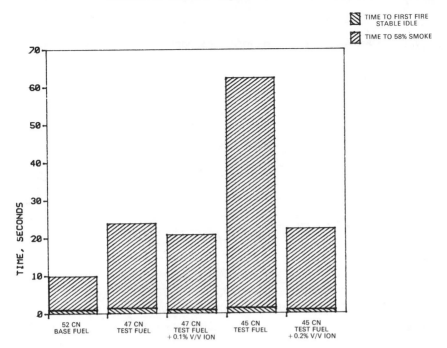

Figure 3.12 The effect of cetane number on cold starting at $-10\,°C$ (1.6-litre indirect injection engine).

However, the use of an electric glow plug starter aid on this engine overcame the potential problems of fuel quality on performance as assessed by time to first fire or to stable idle.

Certain benefits from the use of cetane improvers have also been claimed in terms of reduced emissions of hydrocarbons, carbon monoxide and particulates. The use of iso-octyl nitrate cetane improvers have also been shown to have no effect on engine oxides of nitrogen emissions. This is illustrated by the test data for a 1.6-litre indirect injection diesel engine-powered passenger car shown in Figure 3.13 [15]. The use of 0.1 percent v/v iso-octyl nitrate to upgrade the cetane number of the 47 CN base fuel to 53 CN reduced emissions to a level equivalent to that of the reference (52 CN) diesel fuel.

3.2.4 The future potential of other cetane improvers

Significant interest in the development of alternative cetane improvers has been evident in recent years, and two principal areas of activity have

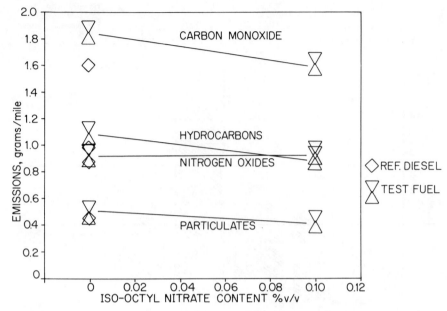

Figure 3.13 Emissions test, US Federal, 1985 (1.6-litre IDI diesel passenger car).

existed. Firstly, attempts have been made to provide additives which can economically compete with iso-octyl nitrate in hydrocarbon fuels. In addition, increasing attention has been focused on the additives required to enable alcohol fuels to be used in unmodified compression-ignition engines.

Research into cetane improvers other than alkyl nitrates has essentially concentrated on nitrate products with specific molecular structures which give enhanced performance. The presence of additional oxygen atoms in

Table 3.11 Cetane improvers other than alkyl nitrates

Compound	Effectiveness relative to iso-octyl nitrate
Iso-octyl nitrate	1.0
2-n-butoxyethyl nitrate	1.2
3-tetrahydrofural nitrate	1.3
4-morphone ethyl nitrate	0.8
Di-tertiary butyl peroxide	0.8
Cyclododecyl nitrate	1.3
5-propyl-1, 2, 3, 4-tetraazacyclo penta-2, 4-diene	1.0
2, 2-di-thiodiisobutyraldehyde	0.8
2-methyl-2-nitro-1-propyl nitrate	1.8

the nitrate compound is a recognized route to improved effectiveness, and this is reflected in the patent literature. Compounds other than nitrates are, however, known cetane improvers, and certain peroxides, tetraazoles and thioaldehydes have been proposed.

Information on the reported performance of certain of these additives relative to iso-octyl nitrate is given in Table 3.11. Increased effectiveness can be achieved with some of these compounds. The majority of the alternatives are, however, complex molecules which cannot be produced at a cost low enough to enable them to compete with iso-octyl nitrate on an economic basis. In addition, certain of the suggested compounds, particularly the peroxide and the thioaldehyde, have unwanted side effects on diesel fuel quality, whilst others, such as the tetraazole, are solids and are not as convenient to use as liquid products.

More apparent success has been associated with the development of additives for use in alcohol fuels. In many parts of the world, methanol and ethanol have received specific attention as substitutes for crude oil-derived fuels. The increasing demand for diesel fuels, particularly in countries with limited crude oil resources, makes the use of alcohol in compression-ignition engines very attractive. The lower molecular weight alcohols which are most readily available are, however, not directly suitable for this application. The use of traditional cetane improvers to improve the auto-ignition characteristics of these fuels is not practical in view of the excessive treatment rates required. Other additives of much greater effectiveness have been identified and their use has been evaluated in countries such as Brazil, South Africa and New Zealand.

The various additives principally considered are polyethylene glycol dinitrates, and reported information summarized in Table 3.12 shows that these are effective at much lower concentrations than iso-octyl nitrate [16]. However, the cost effectiveness of the compounds remains in doubt, and concern over the performance of the treated fuel in low-compression

Table 3.12 Relative performance of dinitrate cetane improvers in ethanol

Additive	Level required to achieve desired ignition delay (% v/v)
Iso-octyl nitrate	16.0
Diethylene glycol dinitrate	5.5
Triethylene glycol dinitrate	4.5
Tetraethylene glycol dinitrate	3.5

engines exists. In addition, the dinitrates are very unstable, and molecules such as diethylene glycol dinitrate can only be used in a desensitized form. The suggested desensitizer is a mix of castor oil and ethylene glycol, comprising about 33 percent of the additive package. Nonetheless, continued interest in this approach to reducing diesel fuel demand exists in many parts of the world.

3.3 Diesel detergents

3.3.1 Diesel engine nozzle coking

In the diesel combustion process fuel vaporization and efficient mixing with available air are essential in ensuring efficient combustion. The fuel injection equipment (pump and injectors) provides the mechanical means of achieving this and its performance is critical in controlling rate of fuel injection and fuel atomization. Optimum performance is only achieved when the fuel injection system is free from deposits and adjusted in accordance with the manufacturer's recommendations. There is, however, a tendency for diesel fuels to form deposits during distribution and use, and these can markedly affect the combustion process.

Critical deposits can form in two basic areas. A build-up of gum or resinous degradation products can occur in the injection system. In severe cases this can result in sticking of pump plungers and injector pintles or needles. Potential problems typically only occur on isolated cylinders, with the resultant misfire, causing loss of power and increased exhaust smoke. Carbon deposits build up on the parts of the injector exposed to hot combustion gases, which can affect both fuel flow and fuel atomization characteristics of the injector. Again, loss in power, increased exhaust smoke and poor starting are the noticeable engine performance problems.

Deposit build-up in fuel pumps and injectors is not a new problem, but in recent years it is becoming more apparent, particularly in areas where fuels containing increased proportions of cracked components are being used. Available information suggests that problems are widespread in the USA and are evident in both direct and indirect injection engines. In Europe the problems are less pronounced, and in the main are associated with indirect injection engines.

Opinions in Europe are that, although cases of slight nozzle coking have been reported with existing fuels in direct injection engines, no performance problems exist. Future fuels are more likely to cause some

Additives influencing diesel fuel combustion

difficulties, and it is possible that certain engines may show increased noise or exhaust emission levels. These could become critical as more stringent legislation is introduced.

The situation concerning indirect injection engines is much more critical. Reservations over current fuel quality exist and serious fouling of nozzles occurs in several engine types. Future changes in fuel composition and quality are expected to further increase the significance of this problem. Mechanical solutions such as oversizing of the nozzle or redesign of the pintle have only been partially successful, and the use of detergent additives to alleviate the problem is being suggested.

Pre-chamber engines with pintle-type injectors are almost exclusively used in passenger car applications because they are generally quieter and smoother running. A typical pintle type injector nozzle is shown in Figure 3.14. An important design feature of this type of nozzle is the pilot fuel discharge during overlap which gives a progressive start to combustion. This helps in promoting early ignition and prevents the excessive rise of cylinder pressure which occurs if too much fuel is released during the initial combustion phase.

Heavy coking in the overlap section can almost eliminate the pilot discharge. This results in noisy engine operation and has been shown to cause increased exhaust emissions. These effects are most evident under low-temperature conditions when ignition delay is greatest.

The potential scope of the problem in this type of engine has been extensively reported in recent years, and test data for a 1.6-litre indirect

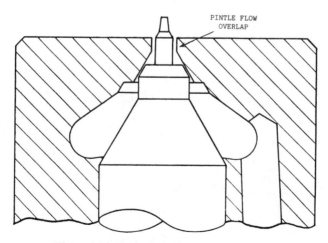

Figure 3.14 Typical pintle-type injector nozzle.

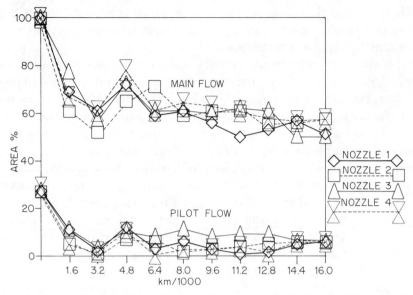

Figure 3.15 The effect of vehicle mileage on injector nozzle fouling (1.6-litre IDI diesel passenger car).

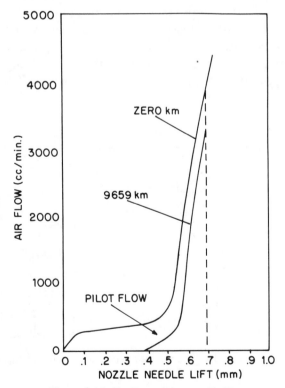

Figure 3.16 Air flow versus needle lift.

Additives influencing diesel fuel combustion

injection engine operated on commercial DERV are considered here [14]. The vehicle was operated on a tape-controlled road simulator under medium-duty driving conditions and at speeds of up to 100 kph. The flow characteristics of the vehicle injector nozzles were determined using an air flow rig conforming to ISO 4010 and monitored at various stages throughout the 16 000-km test programme.

The effects of driving distance on pilot flow and total flow are illustrated in Figure 3.15. The pilot flow, which at start of test represented about 30 percent of the total flow area, was essentially eliminated within 3000 km. Typical flow curves for clean injectors and injectors after 9659 test km are shown in Figure 3.16, further illustrating the loss of pilot flow.

3.3.2 Additives used as diesel detergents

Detergency in diesel engines is generally associated with a range of amine type detergents and polymeric dispersants, and chemical types typically used can be classified as follows:

1. Amines
2. Imidazolines
3. Amides
4. Fatty acid succinimides
5. Polyalkylene succinimides
6. Polyalkylene amines
7. Polyether amines

Additives which are known detergents contain polar groups, and are considered to function by mechanisms which include:

1. Surface action due to the polar group promoting the formation of a barrier film on critical surfaces
2. Dispersant action due to the polymeric additives preventing agglomeration of particulate matter and keeping it dispersed
3. Solvent action due to additives dissolving pre-formed deposits

Surface action and dispersant action are of principal importance to the preventative or keep-clean function of detergents. Solvent action is of much more importance in the removal of existing deposits or clean-up function.

Diesel detergents are marketed in various additive packages offered by several companies, including the major additive manufacturers previously mentioned as suppliers of cetane improvers. In general, the additive packages available appear to be based on compounds which can be classified as polymeric dispersants. The high viscosity of these compounds dictates that they are normally distributed in diluted form, typically 50 percent or more of an aromatic kerosine solvent being used. Diesel detergents have not been extensively used in Europe but increasing attention is being attached to the performance benefits they can offer.

As discussed in the following section, the available products can offer more than the improvements in engine operation achieved by control of injector nozzle fouling. Additional advantages claimed for the products include improved fuel stability, anti-corrosion properties and, in certain instances, combustion improvement. This overall combination of properties makes detergent additive packages attractive in the formulation of the premium diesel fuels now being increasingly offered in markets throughout the world.

3.3.3 Performance benefits from diesel detergents

(a) The effects of reduced nozzle fouling The potential effects of diesel detergents on nozzle fouling and the associated engine performance benefits

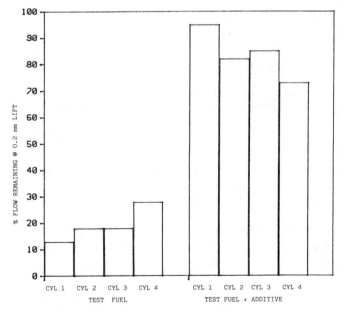

Figure 3.17 The effect of detergent use on nozzle fouling (1.6-litre IDI diesel passenger car).

Additives influencing diesel fuel combustion

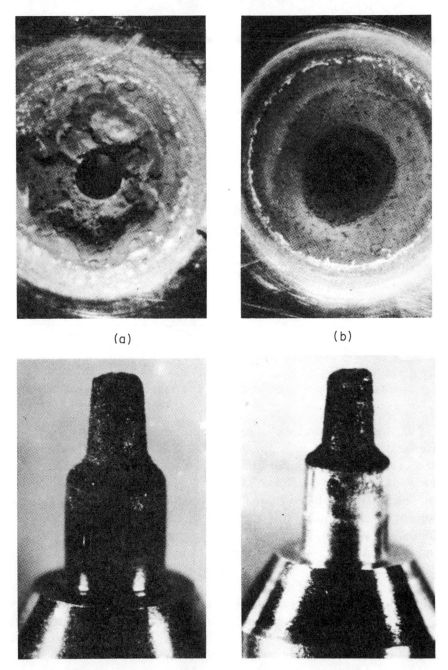

Figure 3.18 Nozzle fouling from fuels with and without detergent. (a) Nozzle from fuel without detergent; (b) nozzle from fuel with detergent.

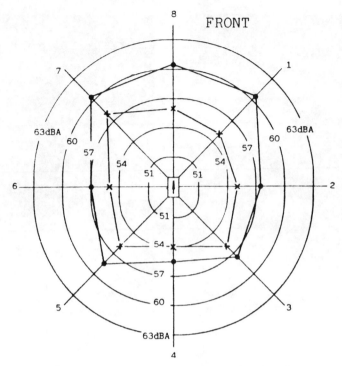

Figure 3.19 All-round noise. Overall 'A' weight noise levels at 7 m distance, engine idling at 850 rpm. • Vehicle fitted with fouled injectors; × vehicle fitted with new injectors.

are illustrated by reference to reported information on a commercially available polyalkenyl succinimide product [14, 15, 17, 18]. The effectiveness of the additive in reducing pintle nozzle fouling was assessed using a 6 h test procedure. As shown in Figure 3.17, test runs completed on the chosen test fuel reduced pilot flow through the injector by almost 80 percent at 0.2 mm needle lift. Equivalent tests completed on the test fuel containing 400 ppm of detergent virtually eliminated fouling. Figure 3.18 shows the state of fouling on nozzles from tests run on fuel with and without detergent.

The reduction in fouling due to additive use has been shown to reduce both noise and engine emissions in many vehicles. Stationary vehicle noise checks for a 1.6-litre indirect injection engine vehicle fitted with clean and fouled injectors are illustrated in Figure 3.19, an increase in noise levels from 55.5 to 60.5 at 7 m from the front of the vehicle being attributed to fouling. Emission tests on a 2.0-litre indirect injection engine vehicle after 1000 km of mixed-duty operation on fuels blended with an

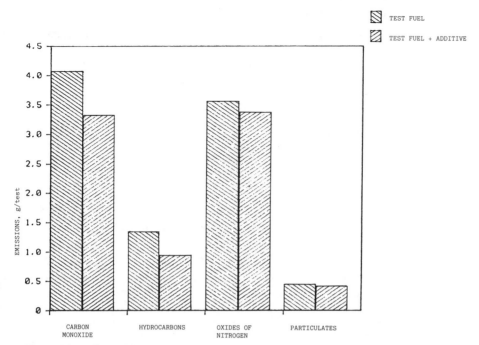

Figure 3.20 Effect of detergent use on vehicle ECE 15 exhaust emissions after 1000 km operation.

without detergent are summarized in Figure 3.20, with lower emissions of carbon monoxide, hydrocarbons and particulates again being attributed to additive use.

It should be noted that certain concern exists on the likely effects of detergents on engines fitted with oversized nozzles. Allowance for the build-up of deposits is made in the design of these injectors and elimination of fouling could prevent optimum operating conditions being achieved.

(b) Other benefits of detergent use An indication of the additional potential benefits of detergent additives in improving fuel stability, fuel corrosivity and combustion efficiency is again provided by reference to reported information on the commercially available polyalkenyl succinimide product.

The effect of this additive on diesel fuel stability is evident from test data completed using the ASTM D4625 procedure, which is based on ageing the fuel at 43 °C. The chosen 13-week test period is considered to correlate with one year's real-life ageing. Results available are summarized in

Table 3.13 The effect of detergent use on gas oil stability

Sample description	Sediment (mg/100 ml)	Adherent gum (mg/100 ml)	Total insolubles (mg/100 ml)
Gas oil ex Africa	0.7	2.0	2.7
+ 100 ppm additive	0.1	0.4	0.5
Gas oil blend with 25% non-hydrotreated LCGO	2.9	4.2	7.1
+ 100 ppm additive	0.5	1.3	1.8
Gas oil ex UK	1.9	0.8	2.7
+ 250 ppm additive	0.6	0.4	1.0

Table 3.13, which indicate that the additive reduces both sediment and gum formation during storage in a wide range of fuels [19]. Similar properties would be expected from other additives, the basic nature of the detergent compounds making them effective in neutralizing the acidic contaminants in diesel fuel which act as precursors in many deposit-forming mechanisms.

Polymeric surfactant materials such as many diesel detergents are also known to have good anti-corrosion properties, and this is very clearly illustrated by Figure 3.21. Comparative results obtained by the ASTM D665 procedure, in which a steel plug is immersed in a mixture of diesel

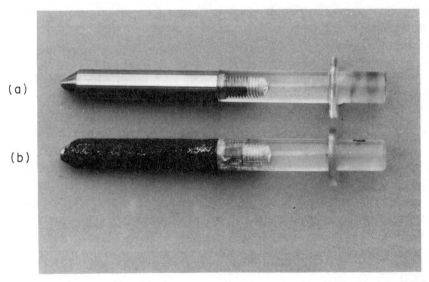

Figure 3.21 The effect of detergent on corrosion as measured by ASTM D665. (a) Fuel with detergent; (b) fuel with no detergent.

Additives influencing diesel fuel combustion

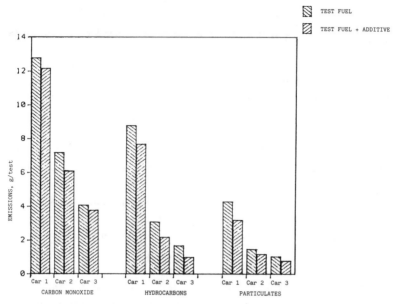

Figure 3.22 Effect of detergent use on exhaust emissions from vehicles with clean injectors (summary of ECE 15 tests on three vehicles).

fuel and water for 24 h at 60 °C, show that the presence of the additive eliminated all corrosion. A recently reported survey of diesel fuel quality in Europe [20] showed certain corrosivity with most diesel fuels tested, although the significance of this in terms of acceptable fuel quality is uncertain.

Potential combustion improvement has also been shown and reduced emissions in additive-treated fuels in an engine with clean nozzles are reported [17]. The information in Figure 3.22 summarizes similar data on other vehicles. Again, reductions in carbon monoxide, hydrocarbon and particulate emissions are evident under standard test conditions.

3.4 Diesel smoke suppressants

3.4.1 Smoke formation in diesel combustion

The presence of smoke and particulates in the exhaust of diesel engines has been a cause of concern for many years. The need for visibly clean exhaust exists, and in many instances this restricts the power available from individual engines. This is particularly true with small direct injection engines and has in part contributed to growth in demand for pre-chamber engine designs. The mechanism of smoke formation is very complex, and is mainly attributed to reactions which occur near full load and when excess fuel is injected [21].

The main causes of in-service emissions of black smoke are poor maintenance of air filters or fuel injectors. The smoke consists primarily of carbon particles or agglomerates of varying size, together with associated condensed fuel and other material. The smoke opacity is dependent on the number and size of the carbon particles present. A smoky exhaust will contain less than 1 percent of the carbon present in the fuel, and certain carbon particles will be found in diesel exhausts under all operating conditions.

Fuel quality can also have an effect on smoke emissions from diesel engines. Changes in fuel density can alter smoke formation simply as a result of changes in mass of fuel injected at equivalent fuel pump settings. Other fuel properties such as chemical composition, essentially aromatic content and certain distillation characteristics have also been shown to affect smoke, but by much more complex mechanisms [22].

The measurement of diesel smoke can be undertaken by various recognized methods which include techniques based on visual comparators, spot tests and opacimeters. The more well-known measurement units are:

1. *Ringelmann number*, determined by visual comparison of smoke with a white chart printed with black grids obscuring 20, 40, 60, 80 and 100 percent of the surface. The scale ranges from 0 (white) to 5 (black).
2. *Bosch number*, determined by passing the exhaust gas through a white filter paper. Deposited carbon particles darken the paper and are taken as a measure of smoke density. Darkening of the filter paper is determined using a reflectometer and values reported on an opacity scale of 0 (clear) to 10 (black).
3. *Hartridge unit*, determined in a smoke meter comprising identical measuring tubes for sampled exhaust and reference air. Light source and photocell are moved from sample tube to reference tube to determine percentage opacity due to exhaust smoke.

Of these procedures, the Bosch and Hartridge techniques are most suited to road vehicle applications and have been extensively used in certification, type approval and surveillance testing of engines. The test results can be considered to correlate with the concentration of carbon in exhaust gas under certain engine conditions and, in the UK, this forms an integral part of the criteria used to assess public perception of exhaust smoke.

Additives influencing diesel fuel combustion

However, as previously discussed, future legislative controls on diesel engine exhaust are expected to be more generally based on specific limitations of particulate mass. It has been widely recognized that smoke represents only the immediately visible fraction of particulate emissions, and that a broader definition which includes all dispersed matter present in exhaust gas at ambient conditions is required. This particulate matter consists of soot particles, condensed hydrocarbons, sulphur-based compounds and other (possibly lubricating oil-derived) material. It is measured as the change in weight of a filter through which cooled exhaust gas has been passed.

In the USA controls on diesel particulate emissions were first introduced on cars in 1982, a limit for new cars of 0.6 g/mile under Federal emission test procedures being imposed. This limit was reduced to 0.2 g/mile in 1986 and in California further reductions to a 0.05 g/mile maximum have been proposed. Limits on particulate emissions from trucks and heavy-duty diesel engines are planned for implementation in 1988.

Within Europe, the Commission of the EEC has recently drafted proposals to limit diesel particulate emissions. The proposed limits for light-duty engines based on the ECE 15 test procedure are 1.1 g/test for type approval and 1.4 g/test for production conformity. These limits are considered to equate to 0.34 g/mile and 0.43 g/mile, respectively, by the Federal test procedure. The limits are planned for introduction in late 1989 for new vehicle designs and in late 1990 for all new vehicles.

3.4.2 Additives which reduce diesel smoke

Various additives exist which are affective in reducing smoke emissions from diesel engines and fuel oil burners. The majority of these are organometallic in nature and compounds containing manganese, iron and barium are generally considered most useful. However, the performance of these additives varies significantly with type of application, and it is additives containing barium which have been shown to be particularly effective in reducing smoke from diesel engines [23].

During the 1960s and 1970s a range of products containing barium, (in some instances, together with calcium or iron) were commercialized as smoke-suppressant additives. These products were very effective in controlling black smoke from diesel engines, and reductions in opacity of up to 50 percent were achieved at additive levels of between 0.25 and 1.00 percent v/v.

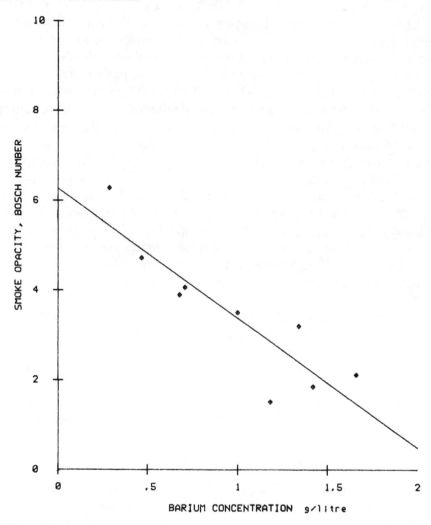

Figure 3.23 Relationship of smoke reduction to barium concentration in a diesel fuel.

The barium content of different additives varied widely, and 25 percent m/m barium was present in certain products. Links between fuel barium content and smoke reduction such as that illustrated in Figure 3.23 were reported [24]. Other workers, however, showed that similar effects could be achieved with products containing very low levels of barium in conjunction with a calcium compound [25].

Many possible mechanisms for the mode of action of smoke reduction by barium-based compounds have been suggested. The most favoured appears to be the inhibiting effect that alkaline earth metals, such as barium, could have on the cracking and dehydrogenation reactions associated with black smoke formation [26]. Despite their obvious effectiveness in reducing smoke from diesel engines, widespread use of smoke-suppressant additives did not develop. Additive cost, together with concern over deposit build-up within engines, the effect of additive use on particulate emissions and the high toxicity of barium compounds, are all considered contributory factors.

Organometallic compounds such as the barium additives used as smoke suppressants leave a deposit after combustion. Thus although, as discussed later, a large proportion of the barium in the fuel is present in the engine exhaust gas, a significant build-up in combustion chamber deposits was typically associated with additive use. In many engines the nature and the location of the deposits were not a cause of concern, and some benefits from additive use were claimed [27]. In other instances problems from additive use had arisen. Engine manufacturers in particular were critical, and were concerned about the effects of increased combustion deposits on piston ring sticking and exhaust valve guttering [28].

Test data have shown that a very high proportion of the metal in smoke-suppressant additives can be exhausted from a diesel engine. Information on barium-based products shows that 85–95 per cent of the metal is emitted as particulates in the exhaust. Thus although use of the additive can considerably reduce carbon-based particulate matter (and hence smoke), no overall reduction in total exhaust particulate was generally observed [29].

The composition of the barium compounds in exhaust gas was also an area of concern in view of the known toxicity of barium. Initial work suggested that, in fuels containing at least 0.3 per cent wt sulphur the majority of barium was as insoluble sulphate. However, doubt concerning the potential impact of barium emissions remained, and was supported by information suggesting strong solubility of barium exhaust particulates in dilute acids.

Research has continued into the means of controlling smoke and particulate emissions from diesel engines, and many potential alternative additives based on transition group and rare earth elements have been identified. However, current specification requirements based on control of particulates essentially precludes the use of organometallic additives, and none are now thought likely to be commercialized.

Interest in organic additives exists, but, in the main, levels of use are extremely high and, in most instances, the suggested products should be considered as fuel compnnents and not additives. Partial or complete substitution of diesel fuel by methanol or methanol is considered by many as a route to reduced particulates, although its long-term efficacy is yet to be established. In other areas short-term objectives are being met by changes in fuel quality to lower sulphur and aromatic contents of the available diesel fuels.

3.4.3 Future control of diesel engine exhaust particulates

The proposed introduction of stringent limits on diesel engine particulate emissions will necessitate the use of alternative technology. In addition to the fuel quality improvements previously discussed, more widespread use of exhaust-treatment devices (e.g. particulate traps) is likely to be required. The use of additives to promote carbon oxidation in these traps may also be necessary.

For many years particulate traps have been seen as a fairly expensive but effective means of controlling diesel engine particulate emissions. The development of suitable filters has been relatively easy, although many problems have been identified with filter blocking and the need for periodic regeneration. Several means of overcoming this problem have, however, been identified and various technically feasible designs of particulate trap now exist. These fulfil both filtration and regeneration requirements by combining temperature-tolerant filter material with an effective means of burning off deposited carbon particulates.

Various forms of these trap oxidizers have been proposed, and include systems based on ceramic monolith and wire mesh filters together with a range of different regeneration techniques [30]. Among these methods is the use of fuel additives to promote continuous filter regeneration.

Many organometallic compounds are known to perform a catalytic function when used in conjunction with filters. Compounds of lead, copper, maganese, iron and cerium have been shown to be effective in promoting carbon oxidation by lowering its ignition temperature. However, concern again exists over the potential impact of using certain of these additives. Nonetheless, there is continued interest in this approach, particularly in systems where additive use can be linked to specific vehicle operation whereas in general fuel production it would not be necessary [31].

3.5 Acknowledgement

The author would like to acknowledge the Associated Octel Company Limited for allowing him to make this contribution and his colleagues at the Engine Laboratory for their assistance in its preparation.

3.6 References

1. R. S. Benson and N. D. Whitehouse, *Internal Combustion Engines*, Pergamon Press, Oxford (1979), pp. 71–5.
2. L. R. C. Lilly (ed., *Diesel Engine Reference Book*, Butterworths, London (1984), pp. 4/1–4/15.
3. United States Bureau of Mines *Ignition Qualities of Hydrocarbons in the Diesel-fuel Boiling Range*, by A. D. Puckett and B. H. Caudle, US Government Printing Office, Washington, DC (1948) (CIC 7474).
4. 'Diesel particulates', *Paramins Post*, Issue 5–2, 10–12, August (1987).
5. American Society for Testing and Materials, *1987 Annual Book of ASTM Standards*; Section 5; Petroleum products, lubricants, and fossil fuels; Vol. 05.01–05.05, American Society for Testing and Materials, Philadelphia (1987).
6. Institute of Petroleum, *Methods for Analysis and Testing of Petroleum and Related Product 1987*, Vols 1 and 2, John Wiley and Sons, Chichester (1987).
7. B. Glavincevski, 'Cetane number estimation of diesel fuels from carbon type structural composition', Society of Automotive Engineers, Warrendale Paper No. 841341 (1984).
8. Associated Octel Company Limited, *European Diesel Fuel Survey, 1985 and 1986*, Associated Octel Company Limited, London (1987) (OPP No. 9).
9. Esso Chemical Research Centre, *Paramins Worldwide Diesel Fuel Quality Survey, 1985*, Esso, Abingdon (1986).
10. United States National Institue of Petroleum and Energy Research, *Diesel Fuel Oils, 1987*, NIPER, Bartlesville (1987), (NIPER-152 PPS) (87/5).
11. Ting-Man Li and R. F. Simmons, 'The action of ignition improvers in diesel fuels', Presented at 21st International Symposium on Combustion, Munich, 1987.
12. Ricardo Consulting Engineers plc, *CF26 Combustion Research Overview*, by A. J. Vokes and H. F. Pettifer, Ricardo, Shoreham (1987) (DP 87/0481).
13. R. M. Olree and D. L. Lenane, 'Diesel combustion cetane number effects', Society of Automotive Engineers, Warrendale, SAE Paper No. 840108 (1984).
14. D. L. Sutton, 'Investigation into diesel operation with changing fuel property', Society of Automotive Engineers, Warrendale, SAE Paper No. 860222 (1986).
15. D. L. Sutton and D. Williams, 'Diesel fuel effects on engine operability and emissions', Presented at 7th World Clean Air Congress, 25–9 August 1986, Sydney, Vol. 4, pp. 320–28.
16. A. J. Schaffer and H. O. Hardenberg, 'Ignition improvers for ethanol fuels',

Society of Automotive Engineers, Warrendale, SAE Paper No. 810249 (1981).
17. X. Montagne, D. Herrier and J. C. Guibet, 'Automotive diesel injector fouling', Society of Automotive Engineers, Warrendale, SAE Paper No. 872118 (1987).
18. D. L. Sutton, M. W. Rush and P. Richards, 'Diesel engine performance and emissions using different fuel/additive combinations', Society of Automotive Engineers, Warrendale, SAE Paper No. 880635 (1988).
19. T. J. Russell, 'The economic benefits of using additives in the production of diesel fuels', Jugoma Professional Publications, Dubrovnik (1987). Presented at Jugoma '87, 28–30 October 1987, Kupari, Dubrovnik.
20. Ethyl Petroleum Additives, *Ethyl European Diesel Fuel Survey, Winter 1986-87*, Ethyl Petroleum Additives, Bracknell.
21. R. Burt and K. A. Troth, 'Influence of fuel properties on diesel exhaust emissions', *Proceedings, Institution of Mechanical Engineers*, **183** (Part 3E), 171–8 (1968–9).
22. G. P. Gross and K. E. Murphy, 'The effect of diesel fuel properties on performance, smoke and emissions' (no publication details) (1976) (American Society of Mechanical Engineers paper ASME 78-DGP-26).
23. K. C. Salooja, 'Burner fuel additives', *Journal of the Institute of Fuel*, January, 37–42 (1972).
24. T. Saito and M. Nabetani, 'Surveying tests of diesel smoke suppression with fuel additives', Society of Automotive Engineers, Warrendale, SAE Paper No. 730170 (1973).
25. C. T. Hare, K. J. Springer, and R. L. Bradow, 'Fuel and additive effects on diesel particulate—development and demonstration of methodology', Society of Automotive Engineers, Warrendale, SAE Paper No. 760130 (1976).
26. C. O. Miller, 'Diesel smoke suppression by fuel additive treatment', Society of Automotive Engineers, New York, SAE Paper No. 670093 (1967).
27. I. Glover, 'The fuel additive approach towards the alleviation of the nuisance of diesel smoke', *Institute of Petroleum Journal*, May, **52**, 137–60 (1966).
28. J. G. Brandes, 'Diesel fuel specification and smoke suppressant additive evaluations', Society of Automotive Engineers, Warrendale, SAE Paper No. 700522 (1970).
29. T. J. Truex, W. R. Pierson, D. E. McKee, M. Shelef and R. E. Baker, 'Effects of barium fuel additive and fuel sulfur level on diesel particulate emissions', *Environmental Science & Technology*, September, **14**, 1121–4 (1980).
30. C. S. Weaver, 'Particulate control technology and particulate standards for heavy duty diesel engines', Society of Automotive Engineers, Warrendale, SAE Paper No. 840174 (1984).
31. G. M. Simon and T. L. Stark, 'Diesel particulate trap regeneration using ceramic wall flow traps, fuel additives, and supplemental electrical igniters', Society of Automotive Engineers, Warrendale, SAE Paper No. 850016 (1985).

4 Diesel fuel additives influencing flow and storage properties

T. R. Coley

5 Wheatcroft Close, Abingdon, Oxon OX14 2BE

4.1	Introduction	105
4.2	**Additives for low-temperature operability**	106
4.2.1	Low-temperature behaviour of diesel fuels	106
4.2.2	Low-temperature properties of diesel fuels	108
	(a) The cloud point test	108
	(b) The pour point test	109
	(c) The cold filter plugging point (CFPP) test	109
4.2.3	Flow-improver additive development	113
4.2.4	The function of flow improvers	113
4.2.5	Fuel response to flow-improver treatment	116
4.2.6	Other wax-modifying additives	122
	(a) Cloud point depressants	122
	(b) Wax anti-settling additives (WASA)	123
4.3	**Other distillate fuel additives**	125
4.3.1	Anti-static additives	125
4.3.2	Antioxidants and metal deactivators	126
4.3.3	Anti-rust and anti-corrosion additives	128
4.3.4	Anti-foam additives	128
4.3.5	Dehazer additives	129
4.3.6	Biocides	130
4.3.7	Lubricity additives	130
4.3.8	Odour masks	131
4.4	**Acknowledgements**	131
4.5	**References**	131

4.1 Introduction

In diesel engines, ignition is achieved by the heat of compression in the combustion chamber rather than by an externally controlled spark, as in a gasoline engine. This fundamental difference between the two principal types of automotive engine necessitates fuels having quite different characteristics and additive requirements.

Unlike gasolines, in which additive use has been common practice for decades, routine use of diesel fuel additives effectively started about 20 years ago with the introduction of cold flow improvers. The purpose of these additives, as the name implies, is to improve the fluidity of the fuel at low temperature when waxes are present. More recently, other types of additive such as antioxidants, detergents and cetane improvers have come into use to maintain or enhance various fuel quality characteristics.

The need for these newer types of additive is due to changes in the petroleum refining industry [1], where more cracking processes are being brought onstream to enable distillate fuel demand to be met from smaller crude runs. In consequence, increasing amounts of cracked components having less than satisfactory ignition or deposit-forming characteristics are finding their way into diesel fuels. At present a relatively small proportion of commercial diesel fuels contain multifunctional additive treatments but wider use is anticipated as the current trend continues.

The additives currently available for diesel fuels fall into two categories: primary treatments for use at the refinery, to meet a specification requirement, and secondary treatments of the 'finished' fuel, by marketers, resellers and end users, for product quality enhancement and also to differentiate from other diesel fuels on the market. The function of these various types of additive and the benefits they can provide to the oil refiner, the fuel marketer and the end user will be discused in the following sections.

4.2 Additives for low-temperature operability

4.2.1 Low-temperature behaviour of diesel fuels

Diesel fuels are 'middle distillates', generally boiling within the range 170–390 °C and comprising a mix of paraffinic, aromatic and olefinic hydrocarbons. They are normally produced by blending two or more refinery streams such as light gas oil (LGO), heavy gas oil (HGO) and kerosine. In a complex refinery with several downstream cracking capabilities, more middle-distillate streams may be available for blending, as Figure 4.1 shows. The proportions of the different components in the finished blend will be determined by their individual characteristics and the requirements of the diesel fuel specification as regards distillation, viscosity, cetane, cold properties, etc.

Typically, 15–30 per cent of the fuel will be paraffinic hydrocarbons. These have only limited solubility and at a critical temperature, when the fuel is cooled, the less soluble, higher molecular weight normal paraffins

Diesel fuel additives influencing flow and storage properties

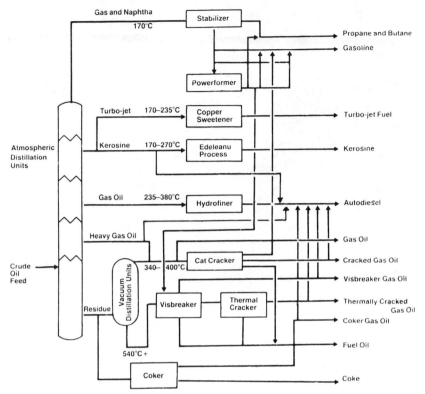

Figure 4.1 Refinery processing. (*Source*: SAE Paper No. 861524).

start to come out of solution as wax crystals. This temperature is known as the cloud point of the fuel.

The wax from a normal, broad boiling range fuel separates in the form of ortho-rhombic crystals with the major axis about 0.5 mm long, as shown in Figure 4.2. These relatively large platelet crystals have a strong affinity for each other and readily interlock to form larger crystal agglomerates. If the fuel is cooled further, more wax will come out of solution until there is sufficient to form an interlocking structure which will prevent the fuel from flowing. The fuel has then reached its pour point, which can occur with as little as 2 per cent wax out of solution [2]. These two physical characteristics of the fuel and the amount of wax coming out of solution at a given temperature will depend on the type of crude oil used, its distillation range and the refining processes from which the blend components were obtained [3].

In a diesel vehicle fuel system the presence of wax crystals can cause

$\longmapsto$ 100 micron

Figure 4.2 Wax in untreated fuel. (*Source*: Exxon Chemical).

operating difficulties by impeding flow through fuel lines and filters. The problems usually occur when starting the engine after overnight shutdown in temperatures below the cloud point of the fuel. If the flow restriction is severe the engine may stall after only a few minutes' running or even fail to start due to fuel starvation. Dewaxing is common practice with lubricant stocks but, due both to the much larger quantities of fuel (and wax) involved and the correspondingly higher costs, it would not be practicable for fuels. Before cold flow improvers became available, avoidance of waxing problems required a cloud point at or below the minimum temperature expected during the winter period. to minimize the likelihood of wax forming in the fuel.

4.2.2 Low-temperature properties of diesel fuels

(a) The cloud point test The cloud point method (IP 219, ASTM D2500) can be applied to any transparent petroleum oils having cloud points below 49 °C. To determine the cloud point a 45 ml sample of the fuel is put into a

Diesel fuel additives influencing flow and storage properties 109

glass jar, cooled at a specified rate and examined periodically. The cloud point is that temperature, expressed as a multiple of 1 °C, at which a cloud or haze of wax crystals appears at the bottom of the test jar. The temperature is measured by a thermometer immersed in the fuel and touching the bottom of the jar.

(b) The pour point test The pour point test (IP 15, ASTM D97) is applicable to any petroleum oil. After preliminary heating to eliminate any effects of its thermal history, a 45 ml sample is cooled at a specified rate and examined, at intervals of 3 °C, for flow characteristics. The pour point is the lowest temperature, expressed as a multiple of 3 °C, at which the fuel is observed to flow. As there is a decreasing temperature gradient from the top to the bottom of the fuel in the pour point test, the thermometer bulb is located just below the surface of the fuel sample.

The cloud and pour points give some information about a diesel fuel, but neither indicates precisely how the fuel will perform in service. Whilst in most instances the cloud point gives too pessimistic an indication of the ability of a diesel fuel to perform satisfactorily, the pour point gives an over-optimistic estimate. The operability limit of a fuel is normally somewhere between those two temperatures.

A test which was developed to correlate more closely with the low-temperature performance of diesel equipment is the Cold Filter Plugging Point (CFPP) test [4]. This is now widely used as the principal test to specify the cold properties of diesel fuels. Some countries also include cloud point or pour point (or even both) in their diesel fuel specifications [5].

(c) The cold filter plugging point (CFPP) test The CFPP test was developed from field data obtained on a fleet of diesel vehicles which were operated on several different quality fuels, some as received from the producing refineries, whilst others had been treated with an additive which interacted with the waxes in a fuel and reduced its pour point. [4]. The vehicles were tested throughout one winter at a location in Sweden, where temperatures suitable for evaluating fuels from different European countries could be expected. Ten base fuels prepared from the various types of crude oil being processed at that period (Middle East, Western Hemisphere, North Africa) were acquired from major oil companies for the field trial. Additive treatment at one or two dose levels gave a total of 25 fuels available for testing in the fleet of 15 diesel vehicles of leading European makes. Cloud points of the fuels ranged from just below zero to

−26 °C, whilst natural or depressed pour points went from −6 °C to below −30 °C.

Selection of the fuel to be used in each vehicle was made following accurate daily predictions by the Swedish Meteorological Institute (SMHI) of the minimum overnight temperature expected at the test location. When the selections had been made, each vehicle fuel tank was drained and half-filled with the new fuel. After fitting a new main filter the engines were run for a time to check for leaks and to purge the previous test fuel from the system. The vehicles were then parked in the open overnight to cool down. Recording instruments were installed to monitor ambient and fuel temperature patterns during the night.

The vehicles were tested the following morning by driving them on the road for a period of 30 min, if possible. Experience had shown that this running time was usually sufficient to enable the engine and fuel system to reach equilibrium temperature and disperse any wax accumulation from the lines and filters. Failures due to fuel starvation caused by wax plugging of the fuel system tended to occur within the first 15 min after start-up. To make the test realistically severe, the vehicles were driven on to the road as soon as possible after engine start-up, without preliminary running at idling speed to warm the engine.

Most of the prepared fuels were tested at some time during the winter, depending on the prevailing temperature conditions, which varied from above zero to below −30 °C. Whenever possible, several tests were run on

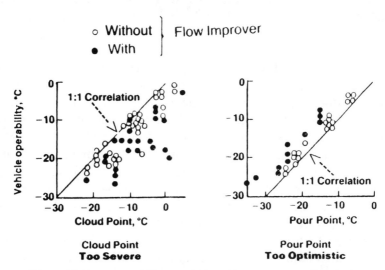

Figure 4.3 Poor operability prediction. (*Source*: Exxon Chemical).

each fuel to determine its operability limit as precisely as possible between the lowest 'Pass' and the highest 'Fail' results. The operability limit was defined as the lowest temperature at which the fuels was able to perform satisfactorily.

Results showed that in all cases the additive-treated fuels were able to perform satisfactorily several degrees lower in temperature than the base fuel without additive. It was also clearly evident that neither cloud point nor pour point was a reliable predictor of cold operability, as Figure 4.3 shows.

Failures were in all cases due to wax plugging of a filter in the fuel system, either the tank strainer, the prefilter at the lift pump inlet or the main filter protecting the injection pump. For this reason, the new test was developed using a filter to assess the ability of the fuel to flow through a vehicle fuel system at low temperatures.

The general arrangement of the cold filter plugging point apparatus is shown in Figure 4.4. It comprises a cooling bath into which a jar containing a small amount of fuel to be tested is placed, a filter holder connected to a 20 ml pipette and a vacuum source. The CFPP method (IP 309, EN 116, BS 6188) applies to diesel fuels and domestic heating oils. A

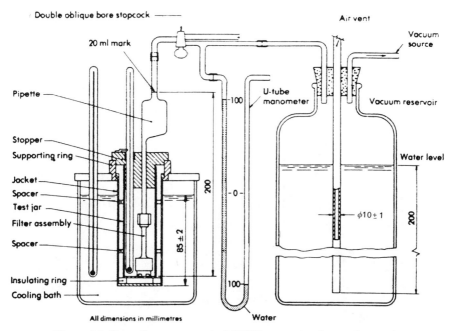

Figure 4.4 General arrangement of CFPP apparatus (manual set-up).

45 ml sample is cooled at a specified rate and examined at intervals of 1 °C for flow through the standardized (45 micron) filter screen immersed in the fuel sample. The CFPP is reported as that temperature (measured near the bottom of the test jar) at which the amount of wax out of solution is sufficient to stop or seriously reduce flow through the filter.

The main test parameters—fuel quantity, cooling rate, vacuum, filter area and porosity—were selected to give a direct correlation with the lowest temperature at which the fuel was able to operate satisfactorily in the vehicle tests. The paper fuel filters used in diesel fuel systems are finer in porosity than that of the CFPP but they also have a much larger surface area to avoid excessive fuel pressure loss. The CFPP screen was selected to have the appropriate sensitivity to fuel wax under the prescribed conditions of the test [6].

Another important requirement was that the test should give an answer in an acceptably short time, to facilitate any necessary adjustment to the CFPP of a batch of fuel in preparation. This was achieved by selecting cooling conditions similar to those of the cloud and pour point tests. The close correlation between vehicle operability limits and the new CFPP test is shown in Figure 4.5. It is also evident that the test can be used for untreated fuels as well as for those containing a flow improver additive.

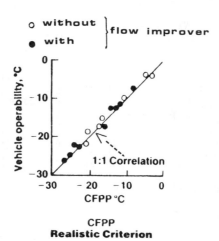

Figure 4.5 CFPP gives best correlation with diesel performance. (*Source*: Exxon Chemical).

Diesel fuel additives influencing flow and storage properties

4.2.3 Flow-improver additive development

Flow-improver additives are ashless polymers which have no influence on the fuel other than its cold-flow properties. Treating levels are typically in the range of 50–500 ppm. The type of material used in those early field tests was an ethylene/vinyl acetate (E/VA) copolymer, and many of the later developments were of the same type or other olefin–ester copolymers. The results of the field tests had demonstrated that wax-modifier additives were effective but the degree of improvement was small and rarely exceeded 5 °C. More potent molecules had to be developed to give greater depression of CFPP and a corresponding improvement in vehicle operability for the additive to be commercially as well as technically viable.

Research in this domain was stimulated by the rapidly evolving situation during the 1970s, when increases in crude oil price changed national policies on energy utilization. Demand for crude oil dropped and many refineries were closed down to help rebalance supply and demand for refined products. Another significant outcome was a heightened concern about fuel economy, which led to a swing in interest to the diesel engine for personal motoring. This increased demand for diesel fuel in a period when refiners were reducing crude runs to avoid overproduction of heavy fuel oil, which was in surplus due to major customers switching to coal [7]. Flow improvers enabled the petroleum industry to increase distillate fuel yield by their use in blends containing higher amounts of heavy components whilst at the same time reducing production of unwanted fuel oil.

4.2.4 The function of flow improvers

As a diesel fuel cools, its solubility for wax decreases and eventually (at the cloud point) it becomes supersaturated and the heavier n-paraffins start to come out of solution as wax crystals. The function of the flow improver is to interact with the wax crystals and modify their growth habit, making them smaller and less prone to form agglomerates which could plug lines and filters in the fuel system. These additives work in two ways; by nucleation and by growth inhibition [3].

Nucleation: the additive is formulated so that, as the fuel cloud point temperature is reached, it creates a large number of nuclei to which the first separating wax molecules attach themselves and form crystals.

Growth inhibition: normal development of the wax crystal will be inhibited by the additive, which adsorbs onto the growing crystal surface,

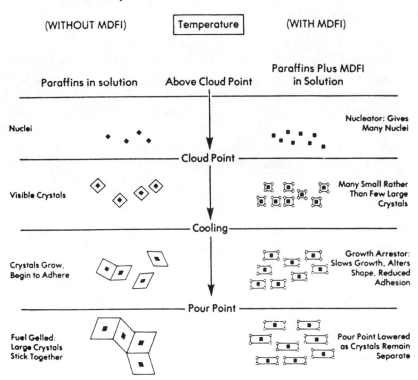

Figure 4.6 Wax crystal growth. (*Source*: Exxon Chemical).

preventing the plate-like growth pattern. The resulting crystals have a more compact shape and will be less prone to agglomerate.

The combined effect of these two additive functions is to promote the formation of a large number of very small crystals rather than relatively few large ones. A diagrammatic representation of the additive function is given in Figure 4.6, whilst photomicrographs of natural and modified wax crystals are shown in Figure 4.7. More detailed information on the additive effect can be seen in Figure 4.8, obtained with a scanning electron microscope (SEM).

These smaller wax crystals are more likely to pass through filter screens or form a permeable coating on the surface of finer filters, enabling the engine to keep running until the fuel has warmed and redissolved the wax. Figure 4.9 illustrates how filters can tolerate larger amounts of additive-modified wax.

Several major oil companies carried out their own field tests to assess the viability of flow improver additives before introducing them into

Diesel fuel additives influencing flow and storage properties

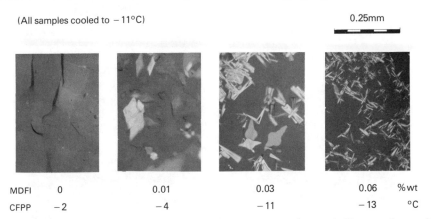

Figure 4.7 Effect of flow improver treat level on wax crystal growth (degree of crystal modification increases with additive content). (*Source*: Exxon Chemical).

Figure 4.8 SEM photograph of wax crystals (after treatment). (*Source*: Exxon Chemical).

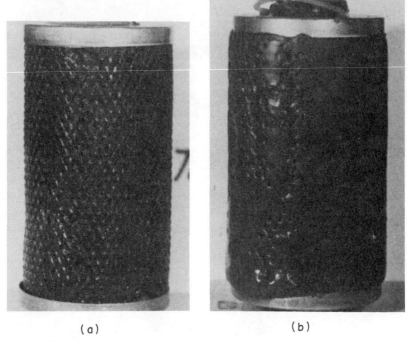

Figure 4.9 Flow improver increases tolerance to wax. (a) Untreated; (b) with flow improver. (Same vehicle performance in both cases.) (*Source*: Exxon Chemical).

commercial fuels. Their data were included in a report published in 1982 by the CEC, after an investigation group had confirmed the continuing relevance of the CFPP for the evolving diesel market [8]. Figure 4.10 presents accumulated results from field tests. Additional data were also obtained by climate chamber testing of diesel vehicles under simulated weather and road-driving conditions and these are given in Figure 4.11, showing a similar correlation between vehicle operability and CFPP predictions.

Practical use over the past 20 years has clearly demonstrated the effectiveness of cold flow improvers, and virtually all European winter-grade diesel fuels (and middle-distillate heating oils) are now treated to attain the relevant CFPP specification.

4.2.5 *Fuel response to flow-improver treatment*

The improvement in CFPP provided by an additive will depend on the characteristics of the fuel as well as on the additive itself. Fuels can vary

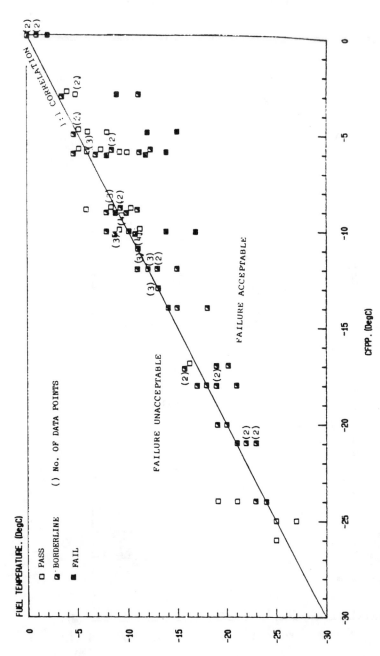

Figure 4.10 Controlled field trials (CFPP versus fuel temperature). *Source:* SAE Paper No. 830596.

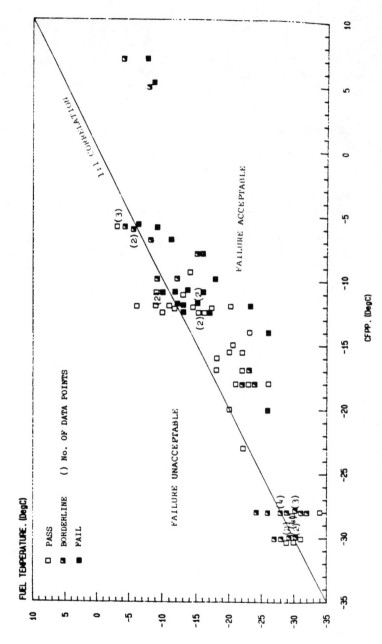

Figure 4.11 Cold-climate chamber tests (CFPP versus fuel temperature).

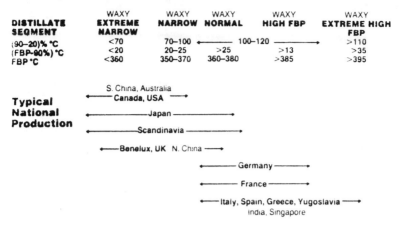

Figure 4.12 Distillate types. (*Source*: SAE Paper No. 861524).

from refinery to refinery and from one country to another, depending on the types of crude being used, the components from which the fuels are blended and the climate in which the fuels are to be marketed. Figure 4.12 indicates how distillates can vary around the world [7].

One of the most important fuel characteristics influencing additive response is its wax content, which will depend primarily on the crude type but also on the distillation range and the sources of the blend components used in the fuel. Streams from cracking processes have lower wax contents than straight-run distillates from the atmospheric unit, as some of the paraffins will have been cracked into smaller molecules for use in gasolines. They also tend to be more aromatic, with better wax solubility characteristics, which can benefit response to additives. On the other hand, complex refineries with many downstream cracking units tend to produce distillates having a narrow boiling range. These are more difficult to treat with flow improver additives than broad-boiling distillates because of the type of wax crystal they produce [3].

Typical response curves are given in Figures 4.13 and 4.14 for different additives and fuel types. These show that response is not continuous but follows the law of diminishing returns, with the depression of CFPP becoming smaller as the treat level increases. To optimize the cost effectiveness of additive use, refiners try to keep treat levels within the range of near-linear improvement in CFPP. The response curves are used to compare the potency of different additives available to refiners, to aid selection of the product best suited to their particular operating patterns.

Diesel specifications include the CFPP as the main criterion of low-

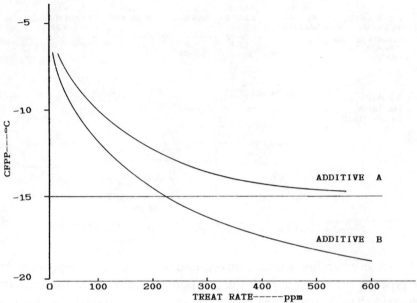

Figure 4.13 Fuel response to different additives. (*Source*: Exxon Chemical).

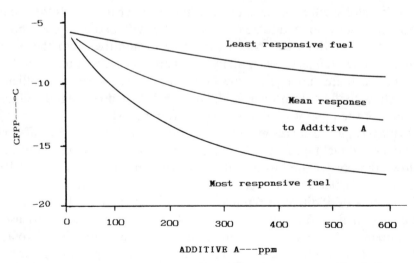

Figure 4.14 Additive effect in different fuels. (*Source*: Exxon Chemical).

Diesel fuel additives influencing flow and storage properties

temperature quality. Control of cloud point level is also required (although not necessarily included in the specification) to ensure that the CFPP relates to how the fuel performs in service. The higher the cloud point is above the CFPP, the greater will be the amount of wax out of solution at the CFPP. If the amount is more than the fuel system can tolerate during the warming-up period, engine failure due to fuel starvation may occur, invalidating the CFPP prediction due to too much wax.

Most commercial diesel fuels treated with flow improver have cloud points no more than 10° or 12 °C above the CFPP specification. Some low-cloud, low-wax fuels for use in extremely cold areas such as northern Scandinavia may have larger spreads but these are exceptional cases.

Before flow improvers became available, the lowest temperature for satisfactory vehicle operation was at or just below the fuel cloud point. Even though only a small amount of wax was out of solution at that temperature, the large wax crystals (Figure 4.2) separating from the fuel would seriously restrict flow through fuel lines or the filters protecting the fuel pump and injectors. The availability of cold flow improver additives has enabled fuels to operate satisfactorily at temperatures well below the fuel cloud point. However, it must be said that flow improver treatment to avoid waxing problems is preventive rather than curative.

Once wax has formed, the use of an additive will not change the waxes already present, although, if the temperature falls lower, it will interact with and modify newly separating waxes. The only way to overcome the problem of flow restriction is to eliminate the wax, either by the use of heat to melt it, by adding a powerful solvent to dissolve it or by physically removing it from the entire fuel system. None of these is easy and certainly cannot provide a quick cure. One further emergency approach might be to run without the offending filter, but this could result in damage to the fuel-injection equipment and possibly invalidate any warranty, so it cannot be recommended.

Guidance given in some diesel vehicle manuals is to add gasoline or kerosine to the diesel fuel tank to help dissolve wax. They are effective wax solvents and will lower the cloud point and the CFPP of the resulting blend by approximately 2 °C for each 10 per cent of added diluent, as indicated in Figure 4.15. However, gasoline must be used with care to avoid the possible fire and explosion hazards, and some oil companies advise against its use. Kerosine is the more suitable diluent but its use by operators for road vehicles is illegal in the UK. Both gasoline and kerosine will reduce the fuel viscosity and density, so should be added only during

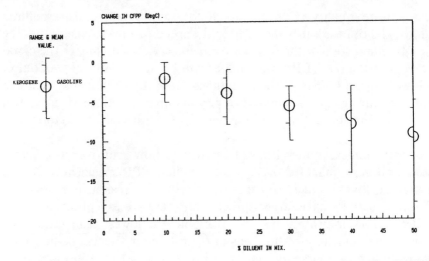

Figure 4.15 Effect of kerosine and gasoline in CFPP. (*Source*: SAE Paper No. 830596).

emergency periods. Regular use would lead to reduced power output and fuel economy and possible wear of injection equipment.

Advice on how to cope with cold weather conditions is given in a booklet published by the British Technical Council [9]. The first recommendation is to use the correct grade of fuel relative to the expected ambient temperature. Oil companies market seasonal grades of diesel fuel to satisfy the climatic needs. If unexpectedly cold spells occur, corrective action may be possible by adding prediluted flow improver to warm, wax-free fuel in refinery or depot storage tanks and checking that the CFPP has been lowered adequately before distributing the fuel to customers.

4.2.6 Other wax-modifying additives

(a) Cloud point depressants A distillate fuel reaches its cloud point when a fall in temperature has reduced its wax solubility, causing some of the heavier n-paraffin components to come out of solution. To depress the cloud point necessitates a change in the fuel solubility characteristics, and this is usually achieved by blending-in a lower cloud point component such as kerosine. As mentioned above, 10 per cent kerosine lowers the cloud point by about 2 °C.

Recent developments in additive technology have identified some olefin–ester copolymers which appear to change fuel solubility so that cloud point depression is possible at the relatively low treat levels associated with additive use, i.e. around 1000 ppm. However, the cloud

Diesel fuel additives influencing flow and storage properties 123

point depression obtainable, even at very high treat rates, was rarely more than three or four degrees, much less than the CFPP improvement usually obtained with flow improvers at treat levels in the range of 100–500 ppm. As these cloud depressants were found to be antagonistic to conventional E/VA flow improvers there is little or no commercial use being made of them in the present state of the art [7].

(b) Wax anti-settling additives (WASA) Additives which inhibit the tendency of wax to settle are a fairly recent development. Conventional cold-flow improvers modify the growth habit of the waxes which separate from solution as the fuel cools below its cloud point. Photomicrograph and scanning electron microscope pictures show that modified waxes are formed as very small, discrete, three-dimensional crystals instead of being large, flat platelets with a strong tendency to agglomerate. As the density of wax is higher than that of the liquid phase, the modified crystals settle more quickly than unmodified crystal agglomerates.

Laboratory and field studies have confirmed this tendency but also showed that, when fuel is drawn from a storage tank, the settled waxes are disturbed by the movement and redistributed in the withdrawn fuel, as illustrated in Figure 4.16. Only if the tank is completely emptied is wax enrichment likely to be significant in some of the last portions of fuel. Under normal circumstances this is unlikely to happen, but it could be a risk if extended spells of cold weather or temperature cycling below the cloud point occur or if the fuel wax content is higher than usual.

- **Potential for inconsistent fuel quality delivered from fuel storage tanks. Final portion of fuel may be wax enriched**

- **Possible crystal agglomeration and/or enlargement under temperature cycling conditions**
- **Incomplete re-solution of wax layer unless temperature rises substantially above fuel cloud point**

Figure 4.16 Effects of wax settling. (*Source*: Exxon Chemical).

The chemistry of this new type of distillate additive, (which may be referred to as WASA) is still proprietary, but its effect is achieved by crystal size reduction to slow down the rate of settling. Stoke's Law on settling of spheres in a fluid states that the rate of settling is a function of

$$\frac{(\text{Diameter})^2 \times \text{density difference}}{\text{Viscosity}}$$

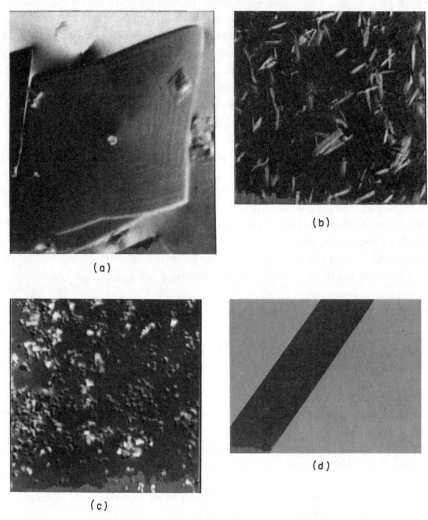

Figure 4.17 A wax anti-settling additive provides crystal size reduction (fuel cooled at 1 °C/h to 10 °C below cloud point). (a) Untreated fuel; (b) normal flow improver; (c) anti-settling additive; (d) human hair (50 microns thick). (*Source*: Exxon Chemical).

Diesel fuel additives influencing flow and storage properties

There is limited scope for controlling density and viscosity but the square-law relationship with diameter means that a five-fold reduction of wax crystal size will slow down settling by one-twenty-fifth, reducing an hourly rate to a daily one. An additional benefit is that the small dispersed crystals redissolve more readily as the temperature rises than does wax which has settled or agglomerated.

The photomicrographs in Figure 4.17 show the smaller crystals obtained with the new additive. Vehicle tests in a climate chamber have confirmed the advantage of the anti-settling additive over a conventional flow improver after an extended period of temperature cycling between the cloud and CFPP levels of the fuel. The fuel with the anti-settling additive was satisfactory but that containing the conventional flow improver failed.

4.3 Other distillate fuel additives

4.3.1 Anti-static additives

Reference has been made earlier (Section 1.3.4) to anti-static additives which are used routinely in aviation kerosine. The risk of building up a charge of static electricity due to high pumping rates pertains to all flammable materials. Some oil companies add anti-static additive to their diesel fuel to give protection during tank refuelling at bulk terminals and depots. Diesel fuels are less volatile than kerosene and normally contain polar compounds which increase conductivity, so there is relatively little risk of excessive charge build-up during filling of engine fuel tanks at service stations. However, some consideration may need to be given to the requirements of trucks having very large-capacity fuel tanks, for which faster filling rates have been proposed.

Test method IP 274/ASTM D2624 covers determination of the electrical conductivity of aviation fuels containing a static dissipator additive. The method normally gives a measurement of the conductivity when the fuel is without an electrical charge. Determination of conductivity is carried out by applying a voltage across two electrodes immersed in the fuel and the resulting current is expressed as a conductivity value. The units used are picosiemens per metre (pS/m). (One pS/m equals $10^{-12}\,\Omega^{-1}$ and replaces the former picomho.) The method is primarily intended as an *in situ* field test for use on product in storage tanks. It may also be used on sampled fuels in the field or the laboratory but care must be taken to avoid contamination.

4.3.2 Antioxidants and metal deactivators

Historically, antioxidant treatment of diesel fuels has not been a requirement, as they were predominantly composed of straight-run components having satisfactory stability characteristics. This is no longer generally true, as the changing fuel demand pattern is obliging oil refiners to incorporate cracked stocks in their diesel pool [10]. Cracking processes are used by refiners to convert heavy streams into lighter components, some of which can be blended into diesel fuel or domestic heating oil. Cracked gas oils are more olefinic and less paraffinic than straight-run gas oils and contain more nitrogen compounds such as pyrroles and indoles. Consequently they have poorer cetane quality and are less stable, being oxidized by free-radical reactions, as in Figure 4.18. Previously they have been blended into heating oil but, with that market declining, it is unable to absorb all the cracked gas oil, so some goes into diesel fuel.

```
INITIATION      RH     ----->  R•                               (1)
                RH + O₂ ----->  R• + HOO•                        (2)

PROPAGATION     R• + O₂ ----->  ROO•                             (3)
                ROO• + RH ----->  ROOH + R•                      (4)
                ROO• + >C=C< ----->  ROO-C-C•                    (5)

TERMINATION     ROO• + ROO•  ⎤   HIGH MOL. WT.
                R•  + ROO•   ⎬   NON-RADICAL                     (6)
                R•  + R•     ⎦   MOLECULES
```

Figure 4.18 The mechanism of product degradation. The process is very complex: principally, free-radical reactions are characterized by initiation, propagation and termination steps. (*Source*: Exxon Chemical).

Diesel antioxidants in current use are mainly hindered phenol or amine structures, used at dose rates of 25–200 ppm. Additive choice and effectiveness will depend on the dominant characteristics of the fuel which determine the complex free-radical reactions that cause product degradation. The manifestations are sediment deposits and colour darkening due to formation of oil-soluble gums.

Additives can improve the fuel stability by acting in two ways: (1) suppressing the normal radical propagation process by reacting with peroxy or alkyl radicals, as in Figure 4.19, and (2) dispersing sediment agglomerates to prevent filter blocking. Although they can reduce the amount of sediment forming, conventional antioxidants tend to be less

Diesel fuel additives influencing flow and storage properties

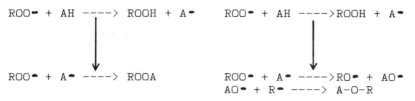

Figure 4.19 The additive effect. *Antioxidant action:* the additive (AH) will suppress the normal radical propagation process by reacting with either peroxy or alkyl radicals. Antioxidants are typically hindered phenol or amine structures. (*Source*: Exxon Chemical.)

effective in controlling colour, so screening tests are needed to identify the most appropriate additive [11].

A stable fuel is one which has a low tendency to form sediment or degrade in colour under normal storage conditions. Most commercial fuels are used within a few weeks, but stocks acquired as strategic reserves may be kept in storage for much longer periods, so it is necessary to have some means of measuring their storage stability.

Refiners need to know whether stabilizer treatment is effective as soon as possible after the fuel is made, so it is necessary to use a predictive test rather than observing how the fuel develops under normal storage conditions over a period of months or even years. However, monitoring of sediment formation and colour change is also necessary for comparison with results derived from predictive tests to indicate what level of confidence can be placed in such techniques.

There is no generally agreed test to predict the stability characteristics of a distillate fuel when stored for prolonged periods at ambient temperature. To achieve this in a relatively short time requires the ageing process to be accelerated by imposing abnormally severe conditions and ASTM D2274 is one of several tests developed for this purpose [11]. A sample of fuel is heated to 95 °C with continuous oxygen bubbling for 16 h, then examined for colour degradation and sediment formation. It is possible that an elevated temperature may induce reactions which would not occur normally, so some laboratories prefer to use a lower temperature for a longer period. Another variant to reduce test severity is presaturation with air or oxygen instead of continuous bubbling.

An alternative test for storage stability is to hold the sample at 43 °C for 100 days. This is believed to correlate with storage at ambient conditions for about one year. However, routine control of product quality obviously requires a rapid test. Specification requirements vary, but a typical sediment limit is around 0.1 mg per 100 ml of fuel.

Metal deactivators reduce oxidation catalysis due to the presence in the fuel of soluble compounds of copper or other metals. Whilst there is less need for these additives in diesel fuel than in gasoline, they may be used in multi-functional packages (at 10–20 ppm) to provide extra protection.

4.3.3 Anti-rust and anti-corrosion additives

Diesel fuels are not naturally aggressive to metals and, because they do not evaporate readily, spillages onto metal surfaces will leave a film which tends to protect against oxidation or corrosion. However, it is almost impossible to avoid the presence of water in diesel fuel systems and anti-rust additives are often used to prevent corrosion or rusting of storage tanks, pipelines and metal fuel system components through which the fuel will pass on its way from the refinery to the engine combustion chamber.

Water can appear in the fuel in various ways, such as by condensation of dissolved water, by incomplete removal after processing operations, by contamination with water used to ballast ship tanks and segregate pipeline products or by 'breathing' of humid air into partially filled tanks due to diurnal temperature variations. Good housekeeping practices will help to minimize the amount of free water and the possible consequences due to its presence in the diesel fuel. Additive treatment, however, may be mandatory in some circumstances such as pipeline distribution or advisable if, for example, a storage tank cannot be drained completely due to distortion of its base.

Rust or corrosion inhibitors are highly surface active, so they readily form a protective layer on the metal surface. The chemical types commonly used for this application are alkenyl succinic acids, esters, dimer acids and amine salts. Treating levels are usually fairly low, in the range of 5–50 ppm by weight. Selection of the additive type and treat rate is usually determined by a rusting test originally developed for turbine oils (IP 135/ASTM D665). In this test a polished steel cylinder is immersed in a stirred mixture of 300 ml fuel and 30 ml water which is maintained at a temperature of 60 °C throughout the test period, usually 24 h. Distilled or synthetic sea water may be used as appropriate and the appearance of the steel cylinder is rated and compared with a control specimen from a portion of untreated fuel.

4.3.4 Anti-foam additives

A recent trend in diesel treatments is the use of anti-foam agents. The foaming tendency of fuels differs, but it is likely to be aggravated by pumping the fuel at relatively high rates to reduce the time required to refuel a vehicle tank at the service station. Although fast refuelling can help the busy motorist, excessive foaming may activate the automatic cut-off in the pump nozzle, thereby reducing the amount of fuel taken or extending the time to fill the tank completely. It may also result in fuel being splashed on the ground and on clothing, as motorists attempt to fill up their tanks.

Environmental issues are becoming more important so, as well as being of direct benefit to the motorist, the use of an anti-foamant can also reduce the risk of oil spills polluting the ground and the atmosphere. Typical anti-foam additives are silicones having a molecular weight suited to the fuel charcteristics. Treat levels rarely need to exceed 20 ppm.

Tests have been developed to assess the tendency of lubricating oils to form a stable foam in an engine, but these are not appropriate for distillate fuels, where the problem is only likely to occur as a result of air entrainment during tank filling. The technique usually employed is vigorous manual agitation to saturate the fuel sample with air bubbles. In a good fuel with low foaming tendency the foam will collapse in less than 10 s, and this is taken as a performance requirement.

4.3.5 Dehazer additives

Distillate fuel can become hazy due to the presence of dispersed droplets of water which have found their way into the fuel during refinery processing and in the distribution system, from the refinery to terminals and depots. The tendency to the formation of hazes or even emulsions will depend on the fuel characteristics (aromatic fuels being directionally more prone to hazing than paraffinic ones) and also on the additives it contains. Good housekeeping practices such as regular draining of storage tanks will help to minimize the amount of free water in the fuel. Dissolved water can appear as a haze when a fuel cools but gross contamination can occur during transportation by ship if adequate de-watering of tanks is not carried out. Normal practice is to allow a few days' settling time, but if the haze persists it may be necessary to resort to additive treatment to accelerate haze clearance.

Additive types which have been found effective as dehazers are quaternary amine salts, used at treat levels in the range 5–50 ppm. Preliminary tests should be made on portions of hazy fuel to determine the appropriate treat level, since overtreatment can sometimes cause haze stabilization. It is necessary to do the preliminary tests on site, as sampled fuels sent to a laboratory for checking usually become clear in the sample container.

It is important to check the effect of any new additive or multifunctional treatment on the water sensitivity of a fuel prior to commercialization of the treated product to minimize the risk of unexpected problems in service. There is no universally accepted test for hazing tendency, but most methods in use involve agitation of sample of the fuel with water, after which it is observed over a period of time to see how long it takes for any water haze to disperse. The test may use either distilled water or synthetic sea water, whichever is the more appropriate. Acceptance standards usually require a distillate fuel to be 'bright and clear' at a given temperature before acceptance.

4.3.6 Biocides

Biocidal treatments are used to control or inhibit growth in storage tank bottoms of bacteria or fungi which can cause blockage of filters if they are drawn out with the fuel. The organisms, which may be aerobic or anerobic, live in the water bottoms and feed at the fuel interface. Anaerobic bacteria derive their oxygen from sulphates in the fuel (sulphate-reducing bacteria) and a by-product of their activity is hydrogen sulphide, which can cause corrosion of steel storage tanks.

Commercial biocides cover a wide range of chemical types, including imines, amines, imidazolines, etc. To be properly effective, the additives have to be soluble in water and also in the fuel. A treat level of 200 ppm is typical for treatment of water bottoms whilst a smaller dose would be injected into the fuel. One major problem with biocides is that the bacteria become resistant, so that the additive type must be changed from time to time.

Biocides are evaluated using a nutrient medium which has been innoculated with the strains of bacteria to be controlled. Treated and untreated samples are incubated for a few days to observe how the growths compare.

4.3.7 Lubricity additives

In cold regions, lubricity additives may be needed to compensate for the lower viscosity of the fuels produced to meet the low-temperature requirements. Their function is to protect against wear of the critical parts of the fuel injection system by forming a protective layer on the metal

Diesel fuel additives influencing flow and storage properties

surfaces. The type of additives used for this application are surface-active agents such as polyfunctional acids and derivatives. Treating levels are usually in the 50–500 ppm range.

Tests for lubricity agents used in distillate fuels are similar to those developed for aviation kerosine, where a stationary object is held in contact with a rotating object immersed in the fuel and the resulting wear scar is measured. The ball-on-cylinder test uses a fixed load and the wear scar on the ball is measured after a predetermined period of time (e.g. 30 min). Another test, on the four-ball machine, involves examination of wear scars produced at different loading levels and determines the maximum load which can be sustained by progressively increasing the load until the lubrication film breaks down and seizure occurs, welding the balls together.

4.3.8 Odour masks

Diesel fuel is less volatile than gasoline, so when spillages occur, the stain and smell will persist for some considerable time. If the diesel fuel spills onto clothing it can be particularly annoying, as the article will probably require dry cleaning to eliminate the traces.

Attitudes to smells are very subjective and vary widely, but odour masks have been used from time to time to improve or eliminate the smell of diesel fuel. There are several odorants available with a choice of scents (fruity, spicy, etc.) for use at treat levels in the range of 10–100 ppm. Indications from odour panel assessments are that the preference tends to be for a neutral rather than a positive odour, so the requirement is for an odour mask rather than a re-odorant.

4.4 Acknowledgements

The author is grateful to former colleagues at the Esso/Exxon Chemical Research Centre for support and advice in the preparation of this chapter and to the Society of Automotive Engineers for permission to reproduce figures from published papers.

4.5. References

1. R. E. Williams, CONCAWE Keynote Presentation, 'Supply and economic considerations', IMechE International Conference, November 1986.
2. C. A. Holder and J. Winkler, 'Wax crystallization from distillate fuels' *J Inst P*, **51**, No. 499, July (1965).

3. J. Zielinski, F. Rossi and A. Stevens, 'Wax and flow in diesel fuels', Society of Automotive Engineers, Warrendale, SAE Paper No. 841352 (1984).
4. T. Coley, L. F. Rutishauser and H. M. Ashton, 'New laboratory test for predicting low-temperature operability of diesel fuels', *JInstP*, **52**, No. 510, June (1966).
5. *Paramins Post*, Issue 1–2, 5, July (1983).
6. CEC Report P-01-74, The cold filter plugging point of distillate fuels. A European test method' (1974).
7. T. R. Coley, F. Rossi, M. G. Taylor and J. E. Chandler, 'Diesel fuel quality and performance additives', Society of Automotive Engineers, Warrendale, SAE Paper No. 861524 (1986).
8. CEC Report P-171-82, 'Low temperature operability of diesels. A report by CEC Investigation Group IGF-3', (1982).
9. BTC Publication, BTC/F1/79, 'Diesel fuel systems for low temperature operation' (1979).
10. T. Coley, 'Diesel trends and fuel additive applications in Europe', *Journal of Japanese Petroleum Institute*, September (1985).
11. R. C. Tupa and C. J. Dorer, 'Gasoline and diesel fuel additives for performance/distribution quality—II', Society of Automotive Engineers, Warrendale, SAE Paper No. 861179 (1979).

5 The use of oxygenates in motor gasolines

G. J. Lang and F. H. Palmer
BP Research Centre, Chertsey Road,
Sunbury-on-Thames, Middlesex TW16 7LN

	Abbreviations used	134
5.1	Introduction	135
5.2	Historical usage of oxygenates and recent developments	135
5.3	Specification characteristics of alcohols and ethers	137
5.4	Manufacture of oxygenates	140
5.4.1	Methanol	140
5.4.2	Ethanol	140
5.4.3	Tertiary butyl alcohol	141
5.4.4	Mixed alcohols	141
5.4.5	Methyl tertiary butyl ether	141
5.4.6	Tertiary-amyl methyl ether	142
5.4.7	Mixed ethers	142
5.5	Laboratory testing of fuels containing oxygenates	143
5.6	Materials compatability	144
5.7	Distribution of fuels containing oxygenates	145
5.8	Vehicle performance with fuels containing oxygenates	147
5.8.1	Road anti-knock performance	147
	(a) Accelerating knock	150
	(b) Constant-speed knock	151
	(c) Etherified spirit knock	151
	(d) Summary of anti-knock performance for oxygenate blends	151
5.8.2	Driveability	152
	(a) Cold-weather driveability performance	152
	(b) Hot-weather driveability performance	156
	(c) Altitude effects	156
5.8.3	Exhaust emissions	156
5.8.4	Fuel economy	159
5.8.5	Inlet system cleanliness	162
5.8.6	Evaporative emissions	163
5.8.7	Intake system icing	165
5.9	Future trends	166
5.9.1	European situation	167

5.10 Conclusions 167

5.11 Acknowledgements 168

5.12 References 168

Abbreviations used

RON	Research Octane Number ⎫ If preceded by 'B'
MON	Motor Octane Number ⎬ denotes blending
RVP	Reid Vapour Pressure ⎭ value
E70 °C	Percentage volume of fuel evaporated at 70 °C when tested by ASTM D86
E100 °C	Percentage volume of fuel evaporated at 100 °C when tested by ASTM D86
E150 °C	Percentage volume of fuel evaporated at 150 °C when tested by ASTM D86

Oxygenate nomenclature

M	Methanol
E	Ethanol
T	Tertiary butyl alcohol (TBA)
MT	Methyl tertiary butyl ether (MTBE)
HC	Hydrocarbon (i.e. without oxygenate)
ES	Etherified spirit
IBA	Iso-butyl alcohol

Values in front of letter denote percentage volume concentration (e.g.) 3M2T-3 per cent volume methanol + 2 per cent volume TBA. Alternatively, this could be written $M_3 T_2$.

Other abbreviations

CO	Carbon monoxide
HC	Hydrocarbon emissions
NOx	Oxides of nitrogen
KLSA	Knock limited spark advance
CEC	Coordinating European Council
EEC	European Economic Community
EFOA	European Fuels Oxygenate Association
CONCAWE	Conservation of Clean Air and Water in Europe (the oil companies' European organization for environmental and health protection)

CCMC	Comité des Constructeurs du Marché Commun
ECE	United Nations—Economic Commission for Europe
CWD	Cold Weather Driveability in terms of driveability demerits. The higher the number, the higher the driveability malfunctions and hence the poorer performance
HFHP	Hot-fuel Handling Performance in terms of driveability malfunctions = RVP (bar) + nE70 per cent vol
WOT	Wide-open throttle
FE	Fuel economy
FID	Flame-ionization detection

5.1 Introduction

With the phasing out of lead additives from gasoline, oxygenated supplements, covering a range of alcohol and ether types, have been recognized as a means of easing the octane burden on refiners, fuel blenders and marketers. However, since these materials are not hydrocarbons their behaviour in terms of blending and vehicle performance is different from hydrocarbon—only gasolines. Furthermore, although many alcohols and ethers have high octane numbers they are not as effective as lead in raising base fuel octane numbers nor do they offer any valve seat protection. Therefore they cannot be regarded as a means of *total* lead replacement. Thus there is a need to evaluate the performance of oxygenates in blends with gasoline in order to define their scope, to establish credibility with existing gasoline specifications and to ensure acceptable market satisfaction.

This chapter has been structured into several sections for easy reference. It describes first, the oxygenate types and their basic specification features followed by manufacture, distribution, laboratory testing, blend specifications and influence on vehicle performance charcteristics. The vehicle performance section has been further subdivided to separate the various features of performance. The effects of oxygenated supplements on each aspect are discussed together with comments on their likely impact in current and future markets and vehicle designs.

5.2 Historical usage of oxygenates and recent developments

The use of oxygenates in motor fuels and the knowledge that they raise octane numbers is not new. Ethanol and methanol, for example, were recognized as octane boosters back in the 1920s and were accredited with giving 'no knock' and 'smoother burning' [1]. Indeed, Alexander Graham Bell, inventor of the telephone, called alcohol gasoline 'a wonderfully clean burning fuel' and Henry Ford said alcohol was the 'fuel of the future'. Because octane numbers in those days were fairly low (i.e. 60–70 RON) the octane boost afforded by oxygenates would have been readily apparent to the motorist. Indeed, older readers will recall the use of Cleveland Discol gasoline in the UK which, for many years, was sold on the basis that it contained alcohol, giving improved performance.

With improvements in refinery practices and more judicious use of lead additives it is fair to say that interest in oxygenates as motor fuel supplements began to wane until the oil crisis of the 1970s. Then they were regarded not so much a means of octane boosting but more as a way of stretching restricted gasoline supplies. Following on from the oil crisis and the subsequent growing pressures to reduce or eliminate lead additives, emphasis on octane number once again re-emerged.

Not surprisingly, crude oil prices influence the economics of oxygenates. In a low crude oil price scenario, oxygenates tend to be less attractive to the fuel blender. Nevertheless, this is just one facet of oxygenate usage and technical need, since octane boosting can sometimes outweigh cost penalties.

Technically, therefore, oxygenates may be regarded as the means of fulfilling three basic needs: extending the gasoline pool; boosting octane values; and providing the refiner with additional blending flexibility to meet ever-increasing demands on quality.

Misuse, lack of knowledge and fear of the unknown has often led to oxygenates being blamed unfairly for the poor market performance of oxygenated gasolines. This is especially true in the USA, where methanol- and ethanol-containing fuels have been cited as a cause of poor vehicle driveability. Sometimes complaints arising through oxygenate misuse during blending and distribution can be justified, but the root cause can often be traced back to a lack of knowledge on the part of the fuel blender rather than of the oxygenates themselves. Water tolerance and hence phase separation can be particularly difficult in this respect. This may be one of the reasons why a high growth rate in the use of MTBE, which is much less critical in this respect, is foreseen in Europe.

Like any other gasoline component, it is necessary to define fully the scope of each oxygenate type, at given concentrations, through blending studies and in-vehicle evaluations in order to achieve market satisfaction and acceptance. Whilst much data have already been generated on the use of oxygenates in gasoline, much more remains to be determined about their performance in future fuel formulations and future engine configurations.

5.3 Specification characteristics of alcohols and ethers

Following considerable discussion between the EEC Member States and organizations representing motor and oil industry interests (CCMC and CONCAWE in Europe), concentration levels have been agreed for most types of oxygenated supplements available in commercial quantities. A list of alcohols and ethers specified in the official *Journal of the European Communities* [2] is reproduced in Table 5.1.

In accordance with Article 1 of the EEC regulations, Member States must permit the addition of organic oxygenates in gasolines up to but not exceeding the limits indicated in column A. However, they may authorize proportions of organic oxygenated compounds higher than these limits on the understanding that, if the limits exceed those values in column B,

Table 5.1 Oxygenated supplements allowed by the EEC Directive [2]

Compounds	A (% vol)	B (% vol)
Methanol, suitable stabilizing agents must be added[a]	3	3
Ethanol, stabilizing agents may be necessary[a]	5	5
Iso-propyl alcohol	5	10
TBA	7	7
Iso-butyl alcohol	7	10
Ethers containing five or more carbon atoms per molecule[a]	10	15
Other organic oxygenates defined in Section 1	7	10
Mixture of any organic oxygenates[b] defined in Section 1	2.5% oxygen weight, not exceeding the individual limits fixed above for each component	3.7% oxygen weight, not exceeding the individual limits fixed above for each component

[a] In accordance with national specifications or, where these do not exist, industrial specifications.
[b] Acetone is authorized up to 0.8 per cent by volume when it is present as a by-product of the manufacture of certain organic oxygenate compounds.
The use of components other than those specified in Section 1 as additives at concentrations below 0.5 per cent in total is not affected by this Directive.

Table 5.2a Properties of typical gasoline and oxygenates (unleaded)

		Typical gasoline C_4-C_{12} mixture	Methanol CH_3OH	Ethanol C_2H_5OH	TBA $(CH_3)_3COH$	MTBE $CH_3OC(CH_3)_3$
Oxygen content	% wt mass	0	49.9	34.7	21.6	18.2
Specific gravity	at 15 °C	0.73–0.76	0.796	0.794	0.791	0.747
Freezing temperature	°C	—	97.8	−117.3	25.6	−108.6
Distillation temperature	°C	32–210	65	78	70–90 commercial	55
Net heat of combustion	Btu/lb	18900	8570	11550	15050	15000
Latent heat of vaporization	Btu/lb	150	506	396	274	144
Stoichiometric air/fuel ratio	:1	14.6	6.45	8.97	11.3	11.8
Water solubility at 20 °C	%	0.1	Infinite	Infinite	Infinite	1.4
RON		95	112	112	117	110
MON		85	91	95	105	101
Azeotrope formation with hydrocarbons		—	Yes	Yes	Yes	No
Toxicity threshold limit value	ppm	About 500	200	1000	250	Not available
Vapour pressure at 38 °C (RVP)	lb	7–15	4.6	2.5	1–1.5	8.0
Flash point	°C	−43	11	13	15	−28
Ignition temperature	°C	371	446	423	470	460

Blending values refer to average properties at the 10 per cent volume blend level in gasoline.

Table 5.2b Test fuels: laboratory data on alcohols and ether compounds assessed

Blending (B) characteristics (unleaded)	Oxygenate type and concentration level (% vol)											
	M_3	M_3T_2	M_3T_7	M_5T_3	T_7	MT_5	MT_{10}	MT_{15}	$M_3T_2MT_5$	E_5	M_3E_2	M_3IPA_2
BRON	136	117	114	120	105	118	118	118	119	135	139	123
BMON	99	97	93	97	90	105	104	104	101	104	103	99
BRVP (lb)	69.5	51	26	38	18.5	11.5	11.5	10.5	31	31.5	50	55
B density	0.77	0.77	0.775	0.785	0.778	0.75	0.751	0.754	0.76	0.79	0.78	0.764
Water tolerance at 0°C % vol	0.05	0.08	0.33	0.17	0.18	—	—	—	—	0.13	0.10	0.10

See 'Abbreviations Used' for definition of nomenclature.

the fuel-dispensing pumps must be suitably marked. These limits officially became mandatory via EEC law from 1 January 1988.

Obviously, it is not possible within the confines of this chapter to review all the combinations of oxygenates, since the number of permutations possible from the EEC list is vast. However, brief details on the oxygenates and ethers reviewed in this paper are given in Tables 5.2a and 5.2b. Properties given in the tables should be regarded as a guide only, since some of them can vary significantly when used in different hydrocarbon base-stocks.

5.4 Manufacture of oxygenates

5.4.1 Methanol

Methanol (methyl alcohol, wood alcohol, CH_3OH) is the first member of the homologous series of saturated alcohols. The first and oldest process for the production of methanol was the destructive distillation of wood, hence 'wood alcohol'. However, methanol is now produced synthetically using natural gas, coal gas, water gas or sewage gas at high pressure and temperature in the presence of metallic catalysts, and can be decribed by the general reaction equations:

$$2H_2O + 2C \rightarrow CH_4 + CO \xrightarrow{O_2} 2CO + 4H_2 \rightarrow 2CH_3OH$$

Direct synthesis from carbon monoxide and hydrogen (the intermediate product above) may also be performed at elevated temperatures and pressures as follows:

$$CO + 2H_2 \rightarrow CH_3OH$$

5.4.2 Ethanol

Ethanol (ethyl alcohol, grain alcohol, C_2H_5OH) is a naturally occurring potable alcohol produced by the fermentation of fruit juices and vegetable matter by airborne yeast spores. Commercially, there are two major manufacturing routes to the production of ethanol, namely 'natural' and 'synthetic'. The natural route involves the fermentation of carbohydrates (i.e. sugars at controlled temperatures by the addition of selected yeasts). These reactions come under the general heading of 'bio-ethanol' and generally follow the equation:

$$C_6H_{12}O_6 \rightarrow 2C_2H_5OH + 2CO_2$$

The use of oxygenates in motor gasolines

'Bio-ethanol' is wet and requires drying before it can be considered as a motor gasoline supplement. The synthetic route generally involves the hydration of ethylene to ethanol, i.e.:

$$C_2H_4 + H_2O \rightarrow C_2H_5OH$$

Ethylene is readily available from steam cracking where the primary use is in the manufacture of polyethylene.

5.4.3 Tertiary butyl alcohol

Tertiary butyl alcohol (TBA, $(CH_3)_3COH$), the most commonly used of the methanol cosolvents, is produced and marketed by Arco Chemicals Inc. in the form of gasoline grade tertiary butyl alcohol (GTBA). The Arco process consists of a controlled oxidation of isobutane to tertiary butyl alcohol and tertiary butyl hydroperoxide (TBHP). The TBHP is reacted with propylene to produce propylene oxide (PO) and additional tertiary butyl alcohol. The ratio of TBA to PO is approximately 2.0–2.5:1, and the general reactions are shown as follows

$$4(CH_3)_2CHCH_3 + 3O_2 \rightarrow 2(CH_3)_3COH + 2(CH_3)_3COOH$$
Isobutane Oxygen TBA TBHP

$$(CH_3)_3COOH + CH_3CHCH_2 \rightarrow (CH_3)_3COH + CH_3CHOCH_2$$
TBHP Propylene TBA PO

The GTBA tends to be rather impure, (ca 95 per cent vol and contains up to 0.81 per cent vol water). However, this product is satisfactory for motor gasoline blending provided that the correct guidelines are followed.

5.4.4 Mixed alcohols

A number of commercial processes produce mixtures of alcohols that are suitable as motor gasoline supplements; e.g. Snamprogetti's MAS (mixture of alcohols superior) and the Union Carbide product 'Ucarnol', both of which are mixtures of methanol and higher alcohols, with the higher alcohols acting as cosolvent for the methanol.

5.4.5 Methyl tertiary butyl ether

Methyl tertiary butyl ether (MTBE, $CH_3OC(CH_3)_3$), is produced by a number of commercial processes. However, each process relies on a final reaction step that consists of reacting methanol with isobutylene. The reaction is reversible and exothermic, i.e.:

$$(CH_3)_2C=CH_2 + CH_3OH \rightarrow (CH_3)_3COCH_3$$
Isobutylene   Methanol

The Arco MTBE process is an extension of their GTBA production route in that the TBA is dehydrated to form isobutylene and the product reacted with methanol to form MTBE as in the above reaction.

The Houdry process uses isobutane as the feedstock, which is dehydrogenated to isobutylene and reacted with methanol. It is also possible to take field butanes and isomerize them to isobutane, and then to dehydrogenate and react with methanol. The MTBE produced by *all* the above routes is suitable for use as a motor gasoline blending component.

5.4.6 Tertiary-amyl methyl ether

Tertiary-amyl methyl ether (TAME, $CH_3OC(CH_3)_2C_2H_5$) is produced commercially by reacting with methanol an isoamylene in which the double bond is on the tertiary carbon atom (i.e. 2 methyl-2-butene or 2 methyl-1-butene). The exothemic reaction is reversible and catalysed by an acidic cation exchange resin, i.e.:

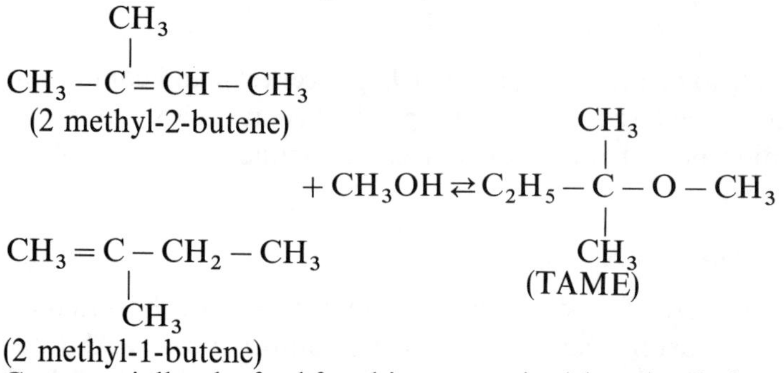

$$CH_3 - \underset{\underset{CH_3}{|}}{C} = CH - CH_3$$
(2 methyl-2-butene)

$$+ CH_3OH \rightleftarrows C_2H_5 - \underset{\underset{CH_3}{|}}{\overset{\overset{CH_3}{|}}{C}} - O - CH_3$$
(TAME)

$$CH_3 = C - CH_2 - CH_3$$
$$\underset{CH_3}{|}$$
(2 methyl-1-butene)

Commercially, the feed for this process is either the C_5 fraction of a light catalytically cracked gasoline stream or a partially hydrogenated light steam-cracked gasoline stream.

5.4.7 Mixed ethers

A number of processes, including the BP 'Etherol' process, use mixed C_4, C_5, C_6 and C_7 olefins derived from catalytically cracked or steam-cracked

The use of oxygenates in motor gasolines 143

streams and etherify the iso-olefins in the presence of methanol to produce mixtures of MTBE, TAME as well as C_6 and C_7 ethers. This process is particularly successful in upgrading the motor octane qualities of cracked naphthas and gives a product known as 'etherified spirits'.

5.5 Laboratory testing of fuels containing oxygenates

Motor gasoline testing in the laboratory has traditionally relied upon a number of limited but well-tried and universally accepted methods of test (i.e. for octane number, density, distillation, stability and vapour pressure). Experience has established that, provided the EEC oxygenate limits are adhered to, the hydrocarbon motor gasoline methods of test are still largely relevant. However, there are a number of exceptions (e.g. hydrocarbon type analysis using fluorescence indicator adsorption (FIA) techniques). These cannot be used because oxygenates affect the interfaces between, for instance, aromatics and olefins and, in addition, they do not fit into any of the hydrocarbon categories. Thus to determine the hydrocarbon types and quantities together with the oxygenates, gas liquid chromatographic methods must be employed. The industry has, as yet, not decided on an exact method of test although plenty of 'in-house' methods are available.

There is also concern over whether to use the wet or dry vapour pressure measurement method, since a lower value is obtained with alcohol blends if water vapour is present. Historically, both the motor and oil industries have used the wet vapour pressure technique (Reid Vapour Pressure, ASTM D323) for specification purposes. This method of test measures the vapour pressure of motor gasoline at 38 °C in the presence of water-saturated air. However, other methods are available in which water vapour is not present.

Water tolerance, that is, the amount of water a motor gasoline can absorb at a given temperature before phase separation occurs, can be determined in the laboratory using ASTM D439. Data from this test give an indication of the water-handling characteristics of motor gasolines in the field (see Section 5.7). The motor gasoline test methods that apply equally to oxygenated and non-oxygenated fuels are given below:

Research Octane Number	ASTM D2699
Motor Octane Number	ASTM D2700
Distillation	ASTM D86
Density	ASTM D1298/D4052

Reid Vapour Pressure   ASTM D323 (but see note above)
Induction Period   ASTM D525
Existent Gum   ASTM D381
Potential Gum   ASTM D873
Lead Content   ASTM D3341/D3237/3229
Bromine Number   ASTM D1159

However, although the test results are correct physically, the results may be misleading in that they may not, for example, reflect the driveability or anti-knock performance of a gasoline in the same way as for a hydrocarbon-only gasoline.

5.6 Materials compatibility

Materials compatibility studies are concerned with metal corrosion, elastomer and plastic attack and the effect on tank linings. Test gasolines containing up to 50 per cent volume aromatic gasoline base stock have been used in one such investigation carried out by the authors' company, together with two reference gasolines (ASTM-C and DIN 51604). In this investigation, corrosion tests of 500 h duration were undertaken at 60 °C and 20 °C using both wet (70 per cent of the blend water tolerance level at 20 °C) and dry fuels. Metal coupons of mazak, zamak (alloys often used in carburettor manufacture) aluminium, brass, mild steel, zinc and terne were tested in oxygenated motor gasolines both with and without corrosion inhibitors.

Elastomer compatibility was evaluated using five commercial mixes of rubbers commonly used in vehicle fuel systems, namely epichlorohydrin, extrusion and moulding grade nitrile, 'Viton' and fluorosilicone. Changes in physical properties were assessed after immersing the elastomers in the oxygenated fuels, a combination of static and cyclic tests at $-20\,°C$, $+20\,°C$ and $+60\,°C$ being used. Tank lining materials were similarly assessed by immersion techniques with coated test coupons.

In general terms, for both corrosion and elastomer compatability it was found that the antagonistic effect of the oxygenated gasolines towards the test specimens showed the order of attack to be

$$\text{methanol} > \text{ethanol} > \text{TBA} > \text{MTBE}$$

as shown in Table 5.3. However, increasing oxygenate content, temperature and, in the corrosion tests, water content increased the magnitude of the attack. The results enable certain recommendations on the most

The use of oxygenates in motor gasolines

Table 5.3 Compatibility of materials with oxygenates

METALS
Corrosion

Resistance ability	*Attack severity*
Descending order of metals	Descending order of fuels
Brass	Methanol
Aluminium	Ethanol
Terne	Tertiary butyl alcohol
Mazak (zamak)	Methyl tertiary butyl ether
Zinc	
Mild steel	

Note: Corrosion rate increased by:
1. Increase in oxygenate content
2. Increase in water content
3. Increase in temperature

ELASTOMERS
Compatibility

Resistance ability	*Attack severity*
Descending order of elastomer	Descending order of fuels
Fluorosilicone	Methanol
'Viton'	Ethanol
Nitrile	Tertiary butyl alcohol
Epichlorohydrin	Methyl tertiary butyl ether

Note: Elastomer incompatibility is increased by:
1. Increase in oxygenate content
2. Increase in temperature
3. Increase in aromatic content

---

judicious selection of component materials to be made (see Table 5.3). Tank linings consisting of zinc silicate, zinc epoxy and polyurethane were considered to give satisfactory performance.

Whilst it is generally recognized that certain alcohols, such as methanol, can display an aggressive nature towards fuel system metals and elastomers, such aggressiveness can be curtailed by the choice of fuel system materials and gasoline additives used.

5.7 Distribution of fuels containing oxygenates

From point of manufacture to point of sale most gasolines have to be transported significant distances by various means, including pipeline, coastal tanker, barge and road tanker, and have been resident in a number of tankage systems, including refinery blending tanks, settling tanks, depot storage tanks and finally the garage forecourt underground tanks. At any point in the distribution system the motor gasoline may (and in practice probably will) come into contact with water. With wholly hydrocarbon gasolines this presents few problems, as hydrocarbons and water are virtually immiscible. However, alcohols are miscible with water, and in the presence of hydrocarbons tend to show a greater affinity for water than for the hydrocarbons. Hence, if a distribution network is too wet, alcohols (and, to a much lesser extent, ethers) tend to be leached out by 'free' water, with two oxygenate-rich phases forming. These may either be in the form of distinct layers or as a hazy or cloudy product. Either way, product quality is adversely affected.

Methanol is infinitely miscible with water and in the presence of hydrocarbons has a low water tolerance, that is, the methanol/hydrocarbon blend cannot retain very much water without separating into two distinct layers. Higher alcohols such as tertiary butanol, isobutanol and isopropanol have high water tolerances and, when added to methanol/hydrocarbon blends, increase the water tolerance of the methanol. Hence higher alcohols of this type are often called cosolvents. Methanol/hydrocarbon blends should only be distributed/marketed with the addition of suitable amounts of cosolvents.

When methanol together with a cosolvent is used in a gasoline, the water bottoms in a storage tank will increase in volume due to some of the methanol transferring from the fuel to the water layer. After a number of successive batches of such gasolines have passed through the tank the water bottoms will start to diminish in volume and will eventually (after several passes, depending upon the amount and type of cosolvent) disappear, since the alcohol content will be high enough to dissolve all the water into the gasoline. The tank should then remain 'dry' provided that no gross water contamination occurs.

There are a number of ways to increase wthe water tolerance of an oxygenated blend, thereby reducing the likelihood of phase separation. These include increasing the aromatic content of the blend, the oxygenate concentration and/or the bulk temperature of the product. It is essential therefore, when distributing oxygenated/hydrocarbon blends (especially alcohols), that the pipe and tankage network should be as water-free as possible. Ethers (e.g. methyl tertiary butyl ether) are soluble in water but only at low levels and exhibit characteristics similar to hydrocarbons rather than alcohols. Consequently, no special handling procedures need be adopted. Water tolerance data for methanol (3 per cent volume), tertiary butanol (7 per cent volume) and methanol (3 per cent volume)

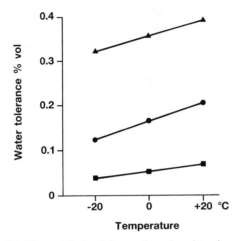

Figure 5.1 Water tolerance for methanol and tertiary butanol.

+ tertiary butanol (7 per cent volume) versus temperature in a 40 per cent volume aromatic hydrocarbon base stock are given in Figure 5.1.

5.8 Vehicle performance with fuels containing oxygenates

The following aspects of vehicle performance can be influenced by the presence of oxygenates in the gasoline:

1. Road anti-knock performance
2. Driveability (hot and cold weather)
3. Exhaust emissions
4. Fuel economy
5. Inlet system detergency
6. Evaporative emissions
7. Intake system icing

5.8.1 Road anti-knock performance

Research and motor octane values (RON and MON) have traditionally been used to control the anti-knock performance of hydrocarbon fuels. Since their inception some 50 years ago questions have arisen over their validity with respect to practical vehicle/engine conditions. With the

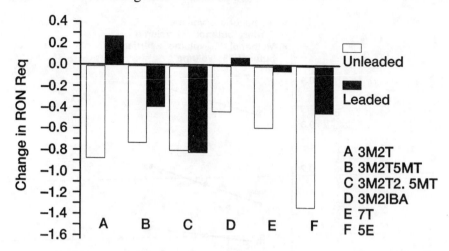

Figure 5.2 WOT accelerating octane requirements at 10 per cent volume olefins. (See 'Abbreviations Used' for identification of oxygenates).

advent of oxygenates the controversy over RON and MON has grown. A number of researchers [3, 4] have found that RON and MON of oxygenated fuels give an unreliable guide to road anti-knock performance, especially for fuels containing ethanol.

Nevertheless, in the absence of suitable alternatives vehicle anti-knock studies have been undertaken using the RON and MON as 'yardsticks', i.e. RON to correlate with low-speed accelerating knock and MON to predict high constant speed knock. In an unpublished study to assess the

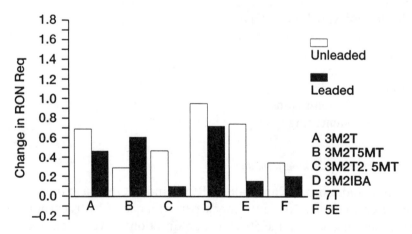

Figure 5.3 WOT accelerating octane requirements at 20 per cent volume olefins.

The use of oxygenates in motor gasolines   149

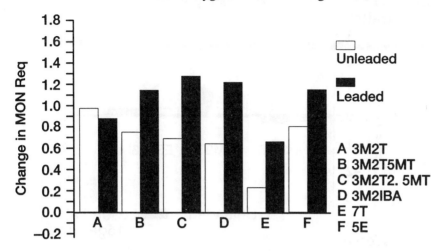

Figure 5.4 WOT constant-speed octane requirements at 10 per cent volume olefins.

true influence of oxygenates all the test fuels were especially prepared at a constant RON and MON (approximately 98/88 for leaded gasoline (0.15 gPb/l) and 95/85 for unleaded) to minimize the number of variables. The results from these road anti-knock tests, given in Figures 5.2–5.5, show the anti-knock performance of gasolines containing various oxygenate types compared with hydrocarbon base fuels containing 10 per cent (volume) and 20 per cent (volume) olefins. From these histograms the following can be deduced.

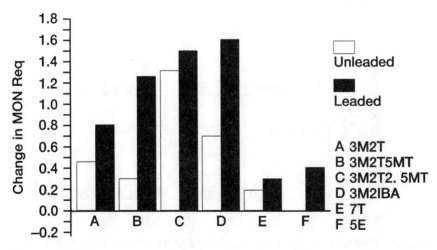

Figure 5.5 WOT constant-speed octane requirements at 20 per cent volume olefins.

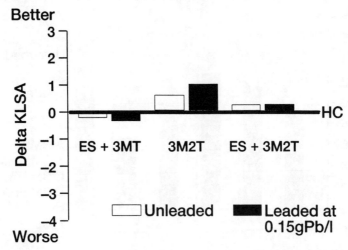

Figure 5.6 Road anti-knock performance of etherified spirit blends under WOT accelerating conditions.

(a) Accelerating knock Virtually all the oxygenated fuels containing 10 per cent olefins (leaded and unleaded) gave a significantly better road anti-knock performance, expressed in terms of RON, compared with hydrocarbon base fuel. In most cases the anti-knock performance of unleaded fuels was better than that of the corresponding blends containing 0.15 gPb/l. These findings are in line with work published earlier [1].

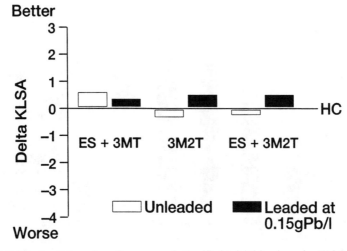

Figure 5.7 Road anti-knock performance of etherified spirit blends under WOT constant-speed conditions.

Anti-knock performance of fuels containing 20 per cent volume olefins was essentially opposite to that of the 10 per cent volume olefin blends, i.e. all fuels gave an inferior anti-knock performance compared with the hydrocarbon base fuels. Also some of the fuels, namely the 7T, 3M2T, 5E and 3M2IBA blends (see Abbreviations used, for explanation of these symbols), gave a worse performance when unleaded. This is somewhat contrary to some earlier findings, and indicates that more work is needed to define more completely the influence of individual types of olefinic compounds and their possible interaction with oxygenated compounds.

(b) Constant-speed knock In the cases of both the high and low olefin level, anti-knock performance was generally inferior with all oxygenate types compared with the hydrocarbon base gasoline. Interestingly enough, the leaded fuels tended to give significantly worse anti-knock performance than their unleaded counterparts. Most fuels gave a similar anti-knock performance at both olefin levels.

(c) Etherified spirit blends The road anti-knock performance of etherified spirit blends produced using components from BP's Etherol process is shown in Figures 5.6 and 5.7. In this case, road anti-knock performance is indicated by knock limited spark advance (KLSA).

On the basis that an increase of two crankshaft degrees KLSA equates roughly to an improvement of one octane number on the road it can reasonably be assumed that the etherified spirit blends, whether they contained 3 per cent volume MTBE or 3 per cent volume MeOH/2 per cent volume TBA, gave anti-knock performances very similar to that of hydrocarbon base fuel under both accelerating and constant speed conditions. However, it must be pointed out that the test cars used in this programme were not identical to the those used to derive the results given in Figures 5.2–5.5, and therefore no direct comparison is valid.

(d) Summary of road anti-knock performance for oxygenate blends In general, it can be said that at low olefin levels the road anti-knock performance of oxygenates are generally equal to or slightly better than that of hydrocarbon only fuels. However, at constant speeds certain oxygenated blends, especially those containing ethanol or methanol, generally give inferior anti-knock performance. However, these deficiencies can be overcome by an adjustment to octane levels. Fortunately, oxygenates have a tendency to give better anti-knock performance in unleaded gasolines, although much depends on the fuel base stock type

and the response of different car models. It is uncertain whether the poorer anti-knock performance in these latest tests (Figures 5.3–5.5) were due to leaner mixture calibrations or just differences in engine types, but earlier work [1] did show progressive improvements in anti-knock performance as MTBE content was increased.

However, single-cylinder research engine studies, carried out by the authors' company, have indicated that oxygenate blends are no better or worse in terms of anti-knock performance when assessed under 'lean-burn' conditions. Further work will be needed in this area to ensure that oxygenates exhibit acceptable anti-knock behaviour in the lean-burn engines of the future.

5.8.2 Driveability

With changes in fuel composition and the move towards leaner-running engines attention is increasingly being focused on vehicle driveability. Not surprisingly, this is increased still further by the use of oxygen-containing compounds in fuels, not only because of their potential contribution to mixture leaning but also because of their impact on volatility characteristics.

Alcohols (and methanol in particular) can have a very large effect on RVP and ASTM D86 distillation curves, as shown in Figures 5.8 and 5.9. Where RVP specifications are limiting it may be necessary to back out butane in order to meet the RVP limit, and this will adversely affect the economics of using alcohols. The distortion of the distillation curves may also make it difficult to meet distillation point specifications when alcohols are present. For these reasons, it is undesirable to simply add these oxygenates to existing gasolines, but they must be introduced as additional components in a blend that has to meet the gasoline specification in every respect.

(a) Cold-weather driveability performance Cold-start and warm-up characteristics have been evaluated on a wide range of oxygenated compounds in blends with gasolines in both chassis dynamometer tests and field trials. Many of these tests have been carried out according to the CEC Cold Weather Driveability (CWD) test procedure [5]. In terms of fuel performance, CWD has traditionally been defined by the E100 °C (percentage volume evaporated at 100 °C using ASTM D86) fuel parameter. The equation, CWD merits = E100 °C – 0.4 E70 °C can also be

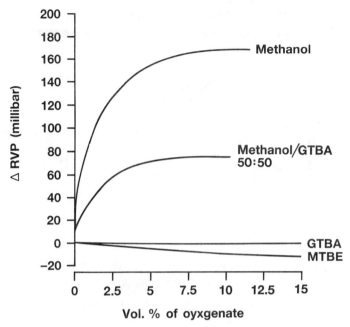

Figure 5.8 Effect of various oxygenates on RVP.

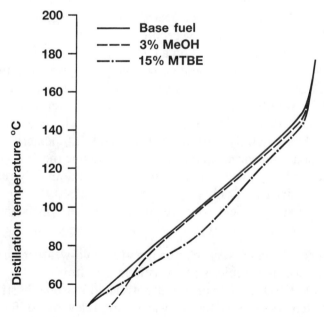

Figure 5.9 ASTM distillation—effect of adding oxygenates.

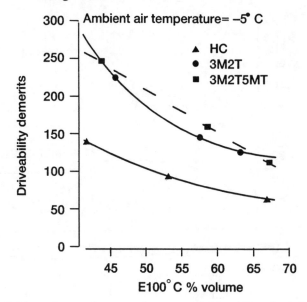

Figure 5.10 Cold-weather performance in terms of driveability demerits (carburetted vehicles).

used. However, more recent investigations [1] have led to the belief that higher distillation points such as E150 °C percent volume are more relevant, especially on fuel-injected cars. This applies generally, irrespective of whether fuels contain oxygenates.

Examples of results obtained on a chassis dynamometer [1] are given in. Figures 5.10 and 5.11. These show that, at a given volatility level, oxygenates generally promote worse cold-weather driveability demerits as judged by stumble/hesitation/surge, etc. and poorer warm-up times. However, such deficiencies can, where necessary, be countered by an increase in fuel volatility. Moreover, if the 200–300 demerit range and 9 minute maximum warm-up time are used to define the limits of performance acceptability, then volatility adjustments are mainly required in the low-volatility range only (say, 45–50 E100 °C).

From cold-climate field trials in Sweden [1] in which ambient air temperatures ranged from 0 to −28 °C it has been shown that:

1. Fuel-injected vehicles gave much better cold-weather driveability performance than vehicles fitted with carburettors
2. Whilst the E100 °C fuel parameter correlated reasonably strongly for carburetted vehicles, E150 °C was more relevant to fuel-injected vehicles

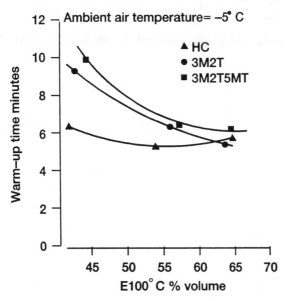

Figure 5.11 Cold-weather performance in terms of warm-up time (carburetted vehicles).

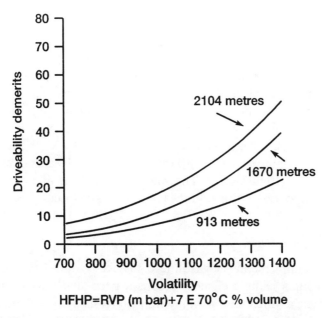

Figure 5.12 Hot-weather high-altitude driveability trials. Effect of altitude on demerits (all fuels excluding 5 per cent volume methanol and without co-solvent).

3. Trends indicated that future vehicles, especially those fitted with fuel-injection equipment, are likely to be more tolerant of low-volatility fuels containing oxygenates

(b) Hot-weather driveability performance Hot-weather driveability performance has been derived using the CEC Hot Weather Driveability (HWD) test method [6] and can be related to gasoline volatility by the hot-fuel handling parameter (HFHP):

HFHP = RVP (mbar) + 7 E70 °C per cent volume

Figure 5.12 shows the relative performance of some oxygenated supplements with respect to gasoline. Because of the inherent nature of hot-weather driveability tests, variations of up to about 30 in terms of HFHP may be regarded as within experimental error. Thus, on this basis, fuels containing 5 per cent volume methanol, without cosolvents, represent about the only type of oxygenate likely to promote a significant deterioration in driveability with respect to wholly hydrocarbon fuels at the same volatility. Such trends have been confirmed in earlier field trials in Italy [7]. In these tests Methanol/TBA blends and, to a lesser extent, MTBE formulations were shown to advantage.

(c) Altitude effects It is well known that altitude affects vehicle driveability under high-temperature conditions and that fuel volatility must be lowered in mountainous regions to ensure acceptable performance [8]. Recent field trials [1] have shown that the use of oxygenates does not require the introduction of any new correction factors or new fuel-volatility parameters. Furthermore, apart from 5 per cent volume methanol blends, all oxygenated fuels gave similar or better hot-fuel handling performance when compared to wholly hydrocarbon fuels of the same volatility (Figure 5.13). The higher latent heat of evaporation of oxygenates may be an important factor in this respect.

5.8.3 Exhaust emissions

Oxygenate supplements in gasoline can affect exhaust emission levels largely through 'leaning effects'. This is a particularly useful feature in 'older' vehicles, which tend to run with progressively richer fuel mixtures with mileage [9]. This is probably due to the fact that older cars were generally set to richer mixture settings and, with general trends in engine

The use of oxygenates in motor gasolines 157

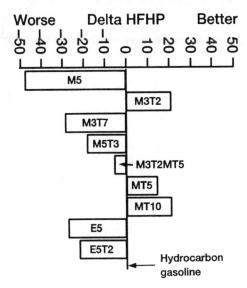

Figure 5.13 Hot-weather driveability. Change in hot-fuel handling parameter (HFHP) to give constant (acceptable) driveability.

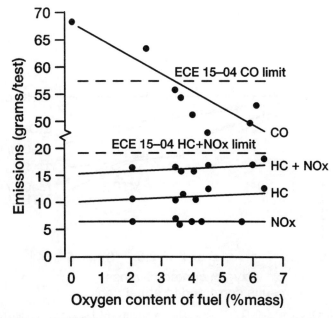

Figure 5.14 Effects of increasing oxygen content—average emissions for five cars (cold start).

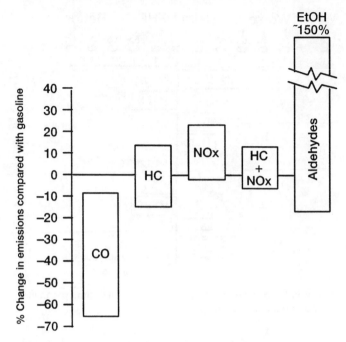

Figure 5.15 Effects of oxygenate blends on exhaust emissions (ECE 15).

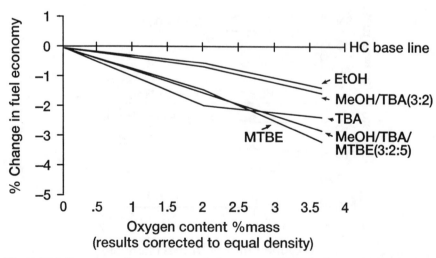

Figure 5.16 Oxygenate effects—average fuel economy (ECE 15 urban cycle) (results from six carburetted vehicles).

The use of oxygenates in motor gasolines 159

tune with mileage, etc, the mixture enrichment that would ensue could be counteracted by the leaning effects of oxygenates. Results in Figures 5.14 and 5.15 show that, irrespective of oxygenate type, CO levels are progressively reduced as oxygen content is increased whilst NOx and HC emissions are barely affected.

Ethanol blends substantially increase aldehyde emissions with increasing blend concentration. However, for all other oxygenate types there is only a marginal tendency for aldehyde emissions to be increased, although such emissions can be corrected by the use of exhaust oxidation or three-way catalysts.

It is interesting to note that the state of Colorado in the USA has mandated the use of oxygenates as a means of improving air quality (i.e. to reduce CO). From 1 January 1988 until 1 March 1988 the minimum mandatory oxygen requirement was 1.5 per cent by weight. This is approximately equivalent to 8 per cent volume MTBE. For the remainder of the programme, from 1 November to 1 March each year, the minimum oxygen content required is 2 per cent by weight or 11 per cent volume MTBE. The oxygen requirements are expected to reduce exhaust tailpipe carbon monoxide emissions by a substantial 24–34 per cent. This, of course, is for cars already fitted with three-way catalyst systems.

5.8.4 Fuel economy

Oxygenated fuels, by definition, contain chemically bonded oxygen in varying amounts, depending on oxygenate type and the amount blended in with gasoline. Also, oxygenates generally have lower calorific values than wholly hydrocarbon fuels. It is quite understandable, therefore, that concern has been expressed about their influence on fuel economy.

Fuel consumption tests have been undertaken on a variety of oxygenated blends, including some containing methanol. In one investigation [1] test fuels were especially prepared to give a range of oxygen contents at constant density. Measurements taken from an average of five cars, when assessed according to the ECE 15 urban cycle, are shown in Figure 5.16 for carburetted cars and Figure 5.17 for fuel-injected ones.

As can be seen, there is a progressive deterioration in fuel economy as oxygen content is increased for TBA, MTBE and the TBA/MTBE/methanol cocktail blends, although the deterioration that occurs with fuels containing up to about 2.5 per cent weight oxygen is considered to be within the precision of the fuel economy test measure-

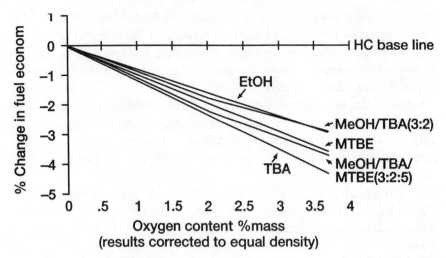

Figure 5.17 Oxygenate effects—average fuel economy (ECE 15 urban cycle) (results from four fuel-injected vehicles).

ments. Beyond 2.5 per cent weight oxygen there is a marked deterioration in fuel economy for all fuels.

In commercial reality, of course, this situation would not arise, since most oxygenate types have higher blending densities than hydrocarbon gasoline. Thus the higher density of commercial blended oxygenated fuels tends to counter potential fuel economy penalties. This is borne out by Figures 5.18 and 5.19, which show that the fuel economy of commercial oxygenated gasolines is essentially unchanged by increasing amounts of oxygenates. Indeed, Figure 5.19 shows that there could be marginal

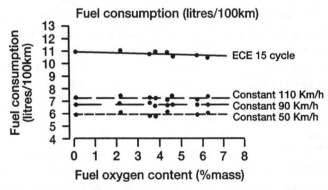

Figure 5.18 Volumetric fuel consumption (average of five vehicles).

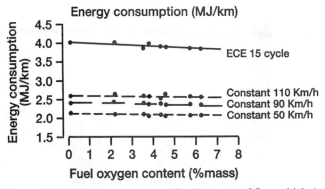

Figure 5.19 Fuel energy consumption (average of five vehicles).

benefits in terms of reduced energy consumption with increased fuel oxygen content.

In a separate investigation [10] fuel-consumption tests were undertaken on a number of cars using gasolines containing etherified spirits from BP's Etherol plant at Vohburg. The fuels were evaluated, with Oxinol and MTBE added, over simulated urban cycles and at low and high constant speeds. As can be seen from Figure 5.20, irrespective of fuel type and operating conditions, no deterioration in fuel economy was observed. Indeed, there was a marginal benefit in favour of the etherified fuels under urban driving conditions. Results obtained on a test car with a closed-loop three-way catalyst car also displayed similar and acceptable

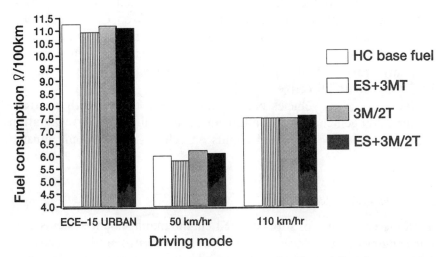

Figure 5.20 Fuel consumption of blends containing etherified spirits (all-car average).

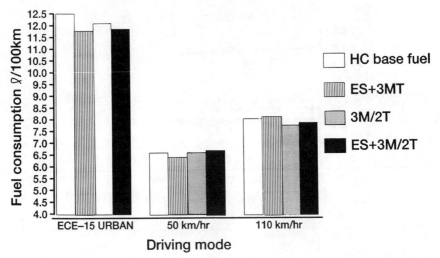

Figure 5.21 Fuel consumption of blends containing etherified spirits (single car fitted with closed-loop three-way catalyst).

fuel economy performance for fuels containing etherified fuels and other oxygenates (Figure 5.21).

From a six-car road programme recently carried out by BP America [10] using fully formulated Oxinol fuels (i.e. blended to meet the required specification and not just with oxygenate added to a finished gasoline) it was found that fuel economy variations between hydrocarbon gasoline and gasoline containing 10 per cent volume Oxinol 50 varied by 0.2 per cent. It was therefore felt that these results confirmed the results generated at BP's Sunbury Research Centre and by others (Arco, Celanese, Dupont and Tennessee Valley Authority): oxygenates, when fully formulated into hydrocarbon gasolines, will not give rise to unreasonable changes in fuel economy provided that concentrations of oxygenates are kept within the EEC Oxygenate Directive.

Nevertheless, as vehicles become more sophisticated with electronic engine management systems and lean-burn engines, further evaluations on the impact of oxygenated blends on vehicle fuel economy must be made.

5.8.5 Inlet system cleanliness

With increasing use of sophisticated engine management systems, lean-burn engines and the need for good emission control and fuel economy, there is a growing need to maintain inlet system cleanliness. The use of

The use of oxygenates in motor gasolines

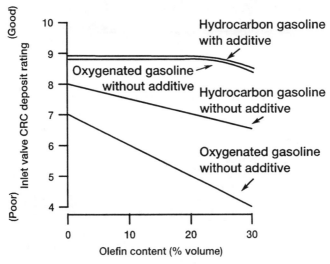

Figure 5.22 Influence of oxygenates and additives on inlet system cleanliness (unleaded gasoline).

oxygenates is inclined to promote greater inlet system deposits, especially in the valve and manifold areas (see Figure 5.22). However, there are suitable gasoline additives available to counter the potential detergency problems with oxygenated fuels. Such additives include the polyisobutene amine types. Provided that they are present at suitable treat levels, the detergency performance of oxygenated fuels can be equal to those of wholly hydrocarbon fuels. The effects of olefin content in Figure 5.22 are intended to represent general trends. Olefin effects depend on the type of olefins present.

Regarding fuel injector fouling, oxygenated fuels are believed to be no worse than hydrocarbon ones, and that suitable additives are available to combat such problems. Indeed, there is some evidence to suggest that the incidence of injector fouling might be slightly reduced by the use of oxygenates, although this has yet to be confirmed.

5.8.6 Evaporative emissions

Growing concern over possible legislation to control fuel evaporative emissions from vehicles has led to the need for a greater understanding of the role of fuel quality. Work done in Europe by CONCAWE [11] has shown that the only fuel parameter to influence evaporative emissions was RVP and, under the ECE conditions of test, the RVP effect is linear.

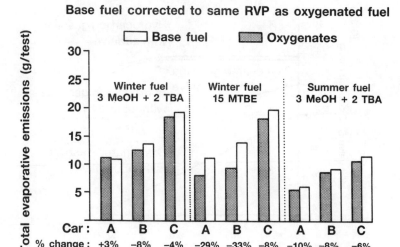

Figure 5.23 Influence of oxygenates on total evaporative emissions base fuel corrected to same RVP as oxygenated fuel.

To assess the influence of oxygenated supplements on evaporative emissions, CONCAWE undertook tests on three different European vehicles using three basic fuels, namely a winter grade containing 15 per cent volume MTBE and winter and summer grades containing 3 per cent volume methanol + 2 per cent volume TBA. For the winter grade fuel, RVP was around 82 kPa, whereas for the summer season fuel it was about 60 kPa.

From the results of the CONCAWE study (Figure 5.23) it can be seen that in most cases the total evaporative emissions are lower for oxygenated fuels than for hydrocarbon ones of the same volatility. The percentage reductions are summarized in Table 5.4. The average reduction in evaporative emissions when 3 per cent volume methanol plus 2 per cent volume TBA is used in both winter and summer fuels is 5.5 per cent

Table 5.4 Percentage change in total evaporative emissions—oxygenated versus HC fuel

Fuel type: Oxygenate: 3 per cent vol	Winter MeOH + 2 per cent vol TBA	Winter 15 per cent MTBE	Summer 3 per cent vol MeOH + 2 per cent vol TBA
Car A	+ 3	− 29	− 10
Car B	− 8	− 33	− 8
Car C	− 4	− 8	− 6

but this is not a significant difference. However, 15 per cent volume MTBE shows a much larger average reduction (23 per cent). This may be significant, but a much larger test programme would be necessary to establish the difference with a high level of confidence.

In summary, it can be said that oxygenated fuels do not increase evaporative emissions, compared with hydrocarbon fuels, provided that they are blended to the same RVP and, in some cases, they may even reduce them.

It is recognized that oxygenated fuel components can influence some analytical test methods since they reduce FID response. Previous work [10] has used a correction factor of 1.05 to account for this effect. It should be stressed that these results were obtained on vehicles without evaporative control systems. There have been suggestions in the USA that alcohols may be preferentially adsorbed in the carbon and not fully desorbed during the purge mode, thus reducing the capacity of the canister. It is conceivable that the canisters used to measure running losses could have been affected, but these had a very large capacity, so it is unlikely that they would have become saturated.

5.8.7 Intake system icing

Intake system icing is caused by ice crystals forming in the fuel-preparation system and adhering to metal surfaces which restricts airflow, causing excessive richness of the mixture. This, in turn, causes either a loss in speed under cruise conditions or repeated stalls at idle, depending on where the ice forms. The conditions inducive to intake system icing are a high level of moisture in the intake system, a low air temperature and a highly volatile fuel. The first two are normally associated with a high relative humidity (> 80 per cent and ambient air temperatures of 3–10 °C i.e. a typical autumn morning).

The addition of alcohols to motor gasoline and, in particular, primary alcohols minimizes intake system icing by reducing the freezing point of the water, hence preventing ice formation. The addition of ethers to motor gasolines will do little to effect the propensity of the fuel to induce intake system icing providing the volatility specification is maintained. However, 'splash' blending of excessive quantities of ethers (i.e. adding ethers 'on top' of fully formulated fuels) may cause intake system icing by greatly increasing the volatility of the motor gasoline. Most modern vehicles are fitted with thermostatically controlled air intake systems to raise the air temperature, which tends to alleviate intake system icing.

5.9 Future trends

As regards future vehicles and fuels, it is of paramount importance to evaluate the performance of oxygenated fuels in preproduction and prototype vehicles to ensure market acceptability. Specific areas of interest include:

1. Closed-loop mixture control systems
2. Electronic fuel injection
3. Lean-burn combustion systems
4. Emission control systems
5. Flexible Fuelled Vehicles (FFVs), i.e. vehicles able to run on any blend of oxygenate and hydrocarbon gasoline

In future studies it is also of paramount importance to maintain a vigil over the validity of existing traditional hydrocarbon specification fuel parameters, bearing in mind their historic origins, to ensure that they continue to predict, accurately, the market performance of hydrocarbon fuels as well as those containing alcohols and ethers.

Various factors will influence the direction, quality and composition of future fuels. The interaction of oxygenates with some of these other factors is illustrated in Figure 5.24. In the longer term it remains to be seen whether alcohols become motor fuels in their own right (e.g. M100 or methanol spiked with hydrocarbons). Methanol fuels are becoming, to some extent, accepted as a means of improving air quality (e.g. lowering tailpipe and evaporative emissions, especially in California). They also enable engine designers to maximize thermal efficiencies. Growth,

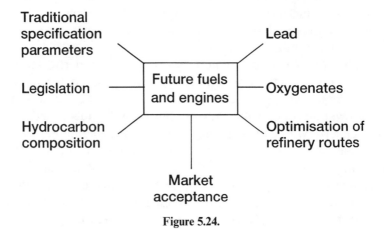

Figure 5.24.

however, will be very dependent on legislative measures and product costs.

5.9.1 European situation

Considerable activity in oxygenated supplements is evident throughout the world, particularly in the USA and Europe, despite the current low crude oil price situation. There is, however, a growing tendency towards a preference for ether-based oxygenated supplements compared with alcohol fuels, for obvious reasons. The European Fuels Oxygenate Association (EFOA), formed in 1984, has completed fuel-blending studies and has already undertaken some vehicle-performance programmes. Concurrently, the EEC has formed a Technical Committee on substitute fuels with a view to assessing the impact of oxygenated supplement usage on the European Community, especially with respect to crude oil savings. An EEC report emanating from the joint efforts of CONCAWE and EFOA was submitted to the EEC TC for approval, and approval was granted in 1988. There is also some inter-oil industry activity to assess the performance of gasolines containing oxygenates in terms of vehicle driveability and anti-knock performance.

5.10 Conclusions

There is little doubt that oxygenated supplements have a role to play in assisting the refinery and fuel blender to meet specification targets and, provided that they have sufficient knowledge, these targets can be successfully achieved. However, market performance must also be considered and meeting specification targets with some oxygenates could be expensive, especially if this means backing out butane to meet front-end volatility requirements. Market penetration is dependent on numerous factors, which include economic, technical and political considerations as well as being influenced by the reduction in the use of leaded gasoline. Performance characteristics discussed in this chapter demonstrate that, within defined limits, oxygenated supplements can give perfectly acceptable vehicle performance. However, full satisfaction can only be achieved if oxygenates are incorporated as part of the initial gasoline formulation, which must comply with existing gasoline specifications. Oxygenate additions after gasoline formulation has been completed (as has been carried out by some independent blenders) are not recommended since they can lead to certain fuel criteria being outside specification limits.

5.11 Acknowledgements

The authors wish to thank the management of the British Petroleum Company plc for giving permission to publish this chapter. They would also like to thank their colleagues at the BPRI Research Centre, Sunbury, and in Britannic House, London, for their invaluable assistance during the construction of the chapter.

5.12 References

1. 'Vehicle performance of gasoline containing oxygenates', Paper No. C319/86, Presented at the The Institution of Mechanical Engineers' Conference on Petroleum Based Fuels and their Automotive Applications, 25–6 November 1986.
2. *Official Journal of the European Communities*, No L 334/22, 12 December 1985, Annex 1.
3. P. D. Histon and R. T. Roles, 'The road anti-knock and pre-ignition characteristics of gasoline containing oxygenates', 5th Alcohol Fuel Symposium, Auckland, New Zealand, 13–18 May 1982.
4. K. Campbell and T. J. Russell, 'The effect on gasoline quality of adding oxygenates', Associated Octel Publication OP82/1, April 1982.
5. The CEC Cold Weather Driveability Test Method, CEC M-09-T-84.
6. The CEC Hot Weather Driveability Test Method, CEC M-08-T-83.
7. F. H. Palmer and A. Tontodonati, 'Road trials to assess the hot weather driveability characteristics of gasolines containing oxygenates in European cars', SAE Fuels and Lubricants Meeting, San Francisco, 1983, SAE Paper No 831706.
8. B. D. Caddock, P. J. Davies, A. W. Evans and R. F. Barker, 'The hot fuel handling performance of European and Japanese cars', Society of Automotive Engineers, Warrendale, SAE Paper No. 780653 (1978).
10. F. H. Palmer, 'The vehicle performance of gasolines containing oxygenated supplements', European Fuels Oxygenates Association (EFOA) Conference, Rome, 22–3 October 1987.
11. J. S. McArragher, W. E. Betts, D. G. Snelgrove, *et al.*, 'Evaporative emissions from modern European vehicles and their control', Society of Automotive Engineers, Warrendale, SAE Paper No. 880315 (1988).

Index

Accelerating knock, 150–151
Alcohols, 13, 21, 31, 87, 137–139
Alkaline metal additives, 59
Alkyl nitrate cetane improvers, 76
Alkyl phenols, 8, 9
Altitude effects, 155, 156
Amides, 91
Amine detergents, 5, 25, 91
Analysis, 35
Anti-corrosion additives, *see* Corrosion inhibitors
Anti-foam additives, 128–129
Anti-haze additives, *see* Dehazers
Anti-icing additives, 19–23
Anti-knock additives, 43–52
Anti-misfire additives, 56
Anti-ORI additives, 52–55
Antioxidants, 3–9, 34, 126
Anti-plug-foulants, 56
Anti-preignition additives, 55
Anti-run-on additives, 55
Anti-rust additives, 128
Anti-sludge additives, 61–62
Anti-smoke additives, 97–102
Anti-static additives, 14, 125
Anti-valve-seat-recession additives, 59, 60
Anti-wear additives, 61, 130–131
Aromatic diamines, 8
Ashless anti-knock additives, 50–51
Aviation mix, 44

Barium additives, 99–102
Biocides, 13–14, 34, 130
Black sludge, 31
Boron additives, 53
Bosch number, 98
Bromine number, 144

Carbon fouling of plugs, 35
Carburettor detergents, 23–26
Carrier oils, 32

CEPP, *see* Cold Filter Plugging Point
Cetane index, 72–73
Cetane number, 69, 71–76
Cetane number improvers, 71–88
CFR engine
  diesel, 71
  gasoline, 43
Cleveland Discol, 136
Cloud point, 75, 108–109, 121
  depressants, 122–123
Co-anti-knock additives, 50
Cold Filter Plugging Point, 75, 109–123
Cold flow improvers, 108, 121
Cold starting, 74, 83–85
Cold weather driveability, 152–155
Combustion
  diesel, 65–66, 83
  gasoline, 40
Combustion chamber deposits, 32, 53–56
Compression ignition, 65
Constant speed knock, 151
Copper in gasoline, 10
Corrosion, 130
  inhibitors, 9, 12, 16–17, 32, 34, 95–97, 128
Cracking processes, 4, 126
Cresyldiphenyl phosphate, 56
Crude oil
  additive response, 119
  price, 2, 136
Cruise icing, 22
Cyclohexyl nitrate, 76, 77

Dehazers, 15–16, 17–18, 129–130
Demulsifiers, 15–16, 17–18
Density, 143
Deposit modifying additives, 53–54
Deposits
  carburettor, 23–24

169

Index

Deposits (*cont.*)
  ice, 19, 22
  induction system, 30
  injector, 27–29
  inlet manifold, 29–30
  inlet port, 29–30
  inlet valve, 29–30
  intake system, 3, 5, 7
  nozzle, 88
  tank, 6
Desensitizers, 88
Detection of surfactants, 35
Detergents, 3
Dewaxing, 108
Dibromoethane, 44, 46
Dichloroethane, 44, 46
Diesel
  detergents, 88–97
  engine design, 67–68
  fuel, 111, 113, 116, 121, 125, 131
  fuel stability, 70, 95–96, 106
  fuel systems, 112
Diethyleneglycol dinitrate, 87
Dinitrates, 76, 87–88
Diolefins, 5
Direct injection, 67
Dispersants, 3, 29, 32, 62
Distillation, 143, 152–154
Distribution, 12, 17, 26
  of oxygenated fuels, 145–147
Drag-reducing agents, 15
Driveability, 4, 28, 31, 57, 58, 152, 156
  additives, 56
Dyes, 12

Elastomer compatibility, 144
Electrical conductivity, 125
Emulsions in gasoline, 15–16, 17–18
Ethanol, 135, 136, 138, 140–141, 144
Ether nitrates, 76
Etherified spirit, 143
  knock, 151
Ethers, 137
Ethylene/vinyl acetate copolymers, 113
Evaporative emissions, 163–165

Exhaust gas, emissions, 2, 3, 70, 76, 83, 86, 95, 156–159
  recirculation, 31
Existent gum, 144
Extended idle icing test, 23

Factory fill additives, 34
Fermentation, 140
Filter screens, 114
Filters, 111, 112, 113, 121, 127, 130
Flame speed, 57
Flexible fuelled vehicles, 166
Flow improver additives, 112–122
Fluidizers, 32
Freeze point depressants, 21
Friction modifiers, 60–61
Front-end octane number, 47
Fuel
  consumption, *see* Fuel economy
  distribution improvement additives, 26–27, 58
  economy, 3, 4, 61, 76, 159–162
  injectors, 27, 88–95, 163
  response to flow improvers, 116–122
Fumarate/vinyl acetate copolymers, 122
Future trends, 166

Gasoline
  distribution, 12
  in diesel fuel, 121–122
  volatility, 19
Glycol dinitrates, 87–88
Grain alcohol, 140
GTBA, 141
Gum, 3–6, 24

Hartridge number, 98
Haze, 15–18, 146
Hot weather driveability, 155–157
HTA (Hydrogenated tallow amines), 58

Icing
  intake system, 19–21, 165

Index

Ignition, delay, 66, 82, 87
  quality, 68–70
Imidazolines, 91
Indirect injection, 68
Induction
  period, *see* Oxidation stability
  system deposits, 7, 29–34
Inhibitor sweetening, 8
Inlet system cleanliness, 162–163
Intake system deposits, 2, 3, 7
Isoamyl nitrate, 76, 77
Isohexyl nitrate, 76, 77
Isooctyl nitrate, 76–79, 83, 86, 87
Isopropyl nitrate, 76, 77

Kerosine in diesel fuel, 121, 122
Knock, 40–41
  diesel, 68

Lead
  additives, 43–48, 59
  alkyls, 43–48, 51, 59
  content, 144
  stabilization, 6
Low temperature operability
  additives, 106–125
Lubricant improvement additives, 60–62
Lubricity additives, 130–131

Manganese, 49
Manifold deposits, 29
Markers, 12
Materials compatibility, 144–145
Metal
  corrosion, 145
  deactivators, 3, 10, 34, 126–128
Methanol, 135, 137–138, 144, 146–151, 158–160
Methyl tertiary butyl ether, 138, 141–142, 144, 145, 153, 158, 161
N-Methylaniline, 51
Misfire, 41–42, 56
Mixed alcohols, 141
Mixed ethers, 142–143
MMT, 49–50
MON, *see* Motor Octane Number

Motor Mix, 44
Motor Octane Number, 46, 134, 143, 146–150
MTBE, *see* Methyl tertiary butyl ether

Noise, 74, 94
Nozzle coking, 88–95
Nucleation, 113

Octane, boosting, 136, 138–139
  requirement, 32
  requirement increase, *see* ORI
Octyl nitrate, 76–79, 83, 95, 86
Odorants, 16
Odour masks, 131
Olefin–ester copolymers, 113
Olefinic compounds, 2, 5, 6, 126, 163
ORI, 32, 52–53
Oxidation, 5, 126
  inhibitors, *see* Antioxidants
  stability, 4–12, 126, 144
Oxinol, 161, 162
Oxygen content of fuel, 137, 157
Oxygenates, 3, 51, 133–168
  manufacture, 140–143
  use in Europe, 167

Particle suspension, 18
Particulates, 3, 70, 97–99, 102
Phase separation, 136, 146
Phosphorus additives, 59
Picket fence rattle, 55
Polybuteneamines, 32, 163
Polyetheramines, 9, 32, 33, 54
Polymeric dispersants, 32–33
Port deposits, 29
Port fuel injector antifoulants, 27–29
Positive crankcase ventilation, 31
Potassium additives, 58
Potential gum, 144
Pour point, 107, 109
Powershield, 59
Preignition, 41–42, 53

R 100 °C, 46, 47

Index

Refinery processes, 107
Reid Vapour Pressure, 134, 143, 144, 152–153, 163
Reodorant, 131
Research Octane Number (RON), 46, 143, 147–150
Ringelmann Number, 98
Road anti-knock performance, 147, 151–152
RON, *see* Research Octane Number
Rumble, 55
Run-on, 41, 42
RVP, *see* Reid Vapour Pressure

Scavengers, 44, 46, 53
Silicone, 129
Smoke emissions, 74, 97–99
  measurement, 98
Smoke-reducing additives, 99–102
Sodium based additives, 59
Solvent oils, 32
Spark aider additives, 57–58
Spark plug anti-foulants, 56
Storage stability, 6, 95, 127
Succinimides, 32, 91, 95
Surface ignition, 55
Surfactants, 13, 15, 16, 62

TAME, *see* Tertiary amyl methyl ether
Tank water bottoms, 8, 9, 13
TBA, *see* Tertiary butyl alcohol
TCP, *see* Tricresyl phosphate
Techron, 54, 55
TEL, *see* Tetraethyl lead
Tertiary amyl methyl ether, 142
Tertiary butyl alcohol, 138, 141, 144, 145, 148–150, 153–155, 157–162
Test methods
  Bromine Number, 144
  Copper Beaker, 11
  Dark Storage, 11
  DB M102E inlet system deposit, 33–34
  Density, 143
  Distillation, 134, 143
  dynamic corrosion, 17, 128

Existent Gum, 11, 144
foaming, 129
Ford Carburettor Cleanliness, 26
icing, 21–23, 131
Induction Period, 10, 144
Lead Content, 144
Lubricity, 131
Motor Octane Number, 143
Multiple Contact Emulsion, 18
Opel Kadett intake system deposit, 26, 33
oxidation stability, 10
Potential Gum, 11, 144
Reid Vapour Pressure, 134, 143, 144
Renault 5 Carburettor Cleanliness, 26
Research Octane Number, 143
rouge suspension, 18
static corrosion, 17
water tolerance, 18, 130
Tetraethyl lead, 6, 44–48
Tetraethyleneglycol dinitrate, 87
Tetramethyl lead, 6, 44–48
TML, *see* Tetramethyl lead
Tricresyl phosphate, 56
Triethyleneglycol dinitrate, 87
Tritolyl phosphate, 59

Unleaded gasoline, 3, 13
Upper cylinder lubricants, 60–61
Urban driving icing test, 22

Valve deposits, 29–34
Valve recession, *see* Valve seat wear
Valve seat wear, 41–42, 48, 59, 60, 135
Volatility, 155

Waring Blender, 18
WASA, *see* Wax anti-settling additives
Water bottoms, 8, 9, 128, 146
Water sensitivity, 18, 146
Water tolerance, *see* Water sensitivity

Wax, 107, 112, 113–123
  anti-settling additives, 123–125
  crystal modifying additives, 113, 122–125
  plugging of filters, 111
  solubility, 122

Wear, 61, 121, 130–131
White smoke, 84
Wild ping, 55

Zinc dialkyldithiophosphate, 59